YOU'RE GOING TO HEAVEN
Whether You Like It Or Not

4-30-19

Dear Deacon Southworth,

May you continue to do God's work

Claire Bertrand

YOU'RE GOING TO HEAVEN
Whether You Like It Or Not

VIRTUAL RELIGION
IN THE
21ST CENTURY

LAURENCE O. MCKINNEY

Copyright © 2018 Laurence O. McKinney
All Rights Reserved

ISBN-13: 978-0-945724-00-1

AMERICAN INSTITUTE FOR MINDFULNESS
32 FOSTER STREET,
ARLINGTON MASSACHUSETTS 02474

GOINGTOHEAVEN.ORG

Contents

Acknowledgments		vii
Prologue		3

Part One: Introduction

1	Preparing for the Coming Faithquake *Faith Tectonics and New World Religions*	13
2	The Metaphysics of Neuroscience *Emerging Perspectives, Shifting Paradigms*	31
3	Painting By Numbers *Virtually Real*	43

Part Two: The Past

4	In the Beginning *From Heaven to Earth*	65
5	Strangers In Paradise: *The Evolution of Chronology*	79

Part Three: The Present

6	Stranded in the Here and Now	101
7	Feel Is How We Real *The Meaning of it All*	113
8	Adventure and Avoidance *Rats, Routines, and Social Media*	131

Part Four: The Future

9	Priests and Prophets *Fulfillment in Real Time*	153
10	Soul Survivors *What Really Happens When We Die*	171

Acknowledgments

NEW DIRECTIONS IN RELIGIOUS THOUGHT always provoke commentary and critique if not controversy. It is especially important, then, to recognize and acknowledge those individuals and institutions that inspired, supported, and enabled a work spanning such a broad divide. In **Neurotheology: Virtual Religion in the 21st Century,** I attempted to name them all, and it's on the website. Lacking a doctorate, nor a spokesman for any religious group or academic institution, how could I attempt a cross-disciplinary work requiring the best of a dozen disciplines and not get lost? The story starts 290 years ago.

In 1727, Harvard was in the midst of a fierce debate as traditional "Old Light" Calvinists fended off "New Light" upstarts such as Jonathan Edwards. Edwards, inspired by natural science at Yale, had read Newton's *Principia* and liked it. So did my first Harvard ancestor, the Rev. William Hobbie. Denied a Harvard appointment for promoting such ideas, he fired back with some angry pamphlets that were soon forgotten. The dispute was not, and in 1807 the "Old Light" faculty decamped to establish Andover Newton Theological Seminary. Harvard's response was to establish the Harvard Divinity

School in 1816, where in 1976, after graduating from both the college and Business School, I found myself taking up the old challenge. Fast forward forty years.

The 200th anniversary celebration at the Harvard Divinity School was in full swing. The School had just been named best divinity school in the country; deans and dons from four graduate schools joined professors, graduates, and guests mingling in the huge white tent. "For one night," announced Dean David Hempton in his lilting Irish accent, "Harvard Divinity will be a party school!" President Drew Faust joined the happy throng of four hundred top tier scholars and practitioners. What a crowd; Harvey (*The Secular City*) Cox explaining a new book, stately Preston Williams taking it in, animated Charles Hallisey responding to ex-students, all three had graded my papers. MacArthur designee Diana Eck (*The Pluralism Project*) greeted friends and colleagues while I got a bear hug from "brother" Cornel (*Race Matters*) West. Isaac Newton appears to be winning. Andover Newton closed that year. Its distinguished faculty has decamped again, this time to Yale.

It was also the 200th anniversary of Henry David Thoreau. Following Thoreau's adage, "Simplify", it would be simplest to acknowledge this entire work is part of a much larger work in progress: the ongoing exploration by my university to illuminate *veritas*, "truth", wherever it lies.

Fortunately for those of us who lived between Harvard and MIT, it seems to be lying all over the place. Like a galactic center, the gravity of that much intelligence in one place draws the finest minds from all over the world. Nationally and internationally recognized experts and innovators in every field amble through Harvard Square like Pokemon Go. Meanwhile, cutting edge advances in dozens of specialties, from theology to molecular biology, everything required for cross-disciplinary research at the highest level, are being dished daily to anyone who has the curiosity at talks, open lectures, symposia, and presentations. Every graduate school, museum, and laboratory adds to the mix. Newton once said "If I have seen further than others, it is because I have stood on the shoulders of giants." With shoulders like those to stand on for forty years, it doesn't take genius to note unusual analogies connecting religious activity, cultural anthropology, developmental neurology and social technology.

Rather than add multiple footnotes and a notes section at the end to formally locate the many references that anchor important

ACKNOWLEDGMENTS ix

points, I tried to include enough specifics in my references to allow anyone with a smart phone to Google the source and learn more at their leisure. In fact a great deal of the information and insight that enlarges the discussions was acquired directly, the result of years of asking questions, some in courses, some in lectures, some at gatherings, some at talks, and some on the street. But what a street! As a teen geek, I had devoured the *Scientific American* cover to cover since I was twelve, giving me a basic fluency in a dozen vocabularies. A month after freshman orientation I was already scribbling notes, captivated by theologian Paul Tillich in an open lecture as he listed the three things we fear the most: death, madness, and the life lived in vain. For those who read the Prologue, it was an epiphany. I wasn't the only one who had questions, and now I was surrounded by answers.

At the same time, Leonard Nash's historic History of Science course was barreling us through every age of science from Ptolemy to relativity by teaching us the prevailing belief and then unveiling the event that upset everything. Just when we can make Ptolemaic predictions work, what's this Copernican thing? His course inspired one of his teaching assistants, the late Thomas S. Kuhn, to write *"The Structure of Scientific Revolutions"* introducing the term "paradigm shift" to the philosophy of science. It also encouraged me to attend countless panels and seminars in the sciences for the rest of my life. In pre-paranoia days one could even drop in on a lecture at Harvard Medical School or attend Grand Rounds at Massachusetts General Hospital in a lab coat.

In Chapter Three, for instance, the discussion of how the brain interprets sight is based on a public lecture David Hubel gave explaining the retina and visual cortex, as well as sidewalk chats with George Wald, who first described rhodopsin. Periodicity, time, and human chronology were covered with Norman Ramsey in a talk in the Science Center. All are Nobel laureates, illustrating the level of expertise available. Prominent Harvard figures such as Lester Grinspoon, George Vaillant and ethnobotanist Richard Evans Schultes contributed to my educational and journal publishing efforts, but I also met Anais Nin, Tim Berners-Lee, Carlos Castenada, and enjoyed E.O. Wilson's offerings since 1975. Prof. Robert Thurmon, head of Columbia's Department of Religion and a Harvard Divinity School graduate, brought the Dalai Lama to teach in college setting for the first time, which explains my tantric initiation with Uma Thurmon, his then teen-age daughter, and a number of others in 1981. As a small part of the Medical School's genetic research, my DNA is in a freezer with Bjork's and the rest of the population of Iceland. It all helped. In all honesty, no single expert in any field could have written this book, but observations collected from scores of some of the best trained experts on the planet made it possible. The parts were all there, waiting to be connected where they seemed to fit.

This book is an update and a revision, but it wasn't spontaneous. Seven individuals had a part in it. The idea began when Manjari Chakravarti, an artist in Santinikitan, near Calcutta, e-mailed how moved she was after reading it. Insights twenty years old from half a world away were still fresh, their powerful uplifting effect undiminished. Then local solar entrepreneur Dave Ellis invited me for dinner to chat neuroscience with his college-age daughter. He called back excitedly to say he'd re-read it. Now he wanted to get a study group together, he hadn't read it carefully before and he was seized by the new perspective. His Catholic faith was unchanged, he said, but this made a big difference. It took a bump out of the road. Claire Bertrand, an educator and nurse, related a similar experience; it bridged a gap and eased her mind. She gave it to a retired Mt. Holyoke professor who was in early Alzheimer's, hoping she could enjoy it, and enjoy it she did. She even asked Claire to thank me for writing it. The words moved and comforted them all. What better reason to bring it back? New research substantiated earlier predictions, and it could be completely updated.

Still, I could not have completed the task without the encouragement and skills of five more individuals. Scholar Priyanka Nandy, with a little assist from Rabindrinath Tagore, reminded me to follow my polar star; my friend Maria Ascher came out of retirement from the Harvard University Press to provide an excellent edit my first version lacked; Pamela Taylor, super copy editor from Texas, snagged everything from misspellings to twenty-two word sentences; and finally publisher Steve Glines, who believe it or not, first collaborated with me on a specialty newspaper for students over 40 years ago. The first four convinced me there were many good reasons to bring it out and spread the word, the others helped me turn it into this book. Thank you all.

YOU'RE GOING TO HEAVEN
Whether You Like It Or Not

Prologue

I was always different. I never had a casual thought in my life. Even as a child I pondered the mysteries of life. As far back as I can remember, I wondered about death. I would look up at the great amber hanging lanterns in the church during Sunday services and wonder what would happen if one dropped on me. I almost wished it would, just to find out. It seemed that nobody had the answer. I went to confirmation class because I thought Sally Meneely was cute; by then I had other things on my mind. Still, starting from about the age of nine, I would return often to a mystery that gradually became the focus of my intellectual life. What was it really? What actually happened?

My father was a Harvard grad who never missed a Yale game. As an editor of the *Harvard Lampoon,* he wrote light verse and later oversaw the family firm, handling management while his vice-president oversaw engineering. The fine sand of the Hudson River attracted foundries that would bake it into intricate molds for cast iron. Founded as a stove works in 1857, "McKinney & Mann," soon reincarnated as Albany Architectural Ironworks, won renown for fancy store fronts in the 1880's, and assumed its third life as James

McKinney & Son when my grandfather entered the firm. My father was born in 1891; I was born in his 54th year, the son of the son of the son at James McKinney & Son. Once "the Tiffany of Ironworks", it now erected steel frameworks.

In 1943, he married my mother, the 27-year-old daughter of a friend, the editor of the *Troy Record*. She'd dropped out of Swarthmore, graduated from Royal Academy of Dramatic Art in London, and wrote advertising copy. He had written a Hasty Pudding show, was Albany's major culture maven, never made much money, and died at seventy-seven of Hodgkin's disease. He was delighted when Harvard accepted me the year of his 50th reunion and I was still there, at graduate school, when he died. They named the library of the Albany Institute of History and Art after him and ran obituaries and editorials for a week. (*McKinneyLibrary.org*)

When I was a child, I looked forward to the evenings after he came home from "the plant". I once thought he worked with vegetables and not at an office. Since my mother said he "made money" there, I assumed they used long strips of copper with a penny die stamp. Every night, he would answer any three questions we had - anything at all. "Where does paint get its color?" "From pigments, in a carrier base." I imagined colored pigs frolicking on the decks of aircraft carriers, safe at home in their naval base.

He always had the answers.

Every spring the carnival came to town. James E. Strates Shows would arrive and pitch its tents in a huge field at the bottom of the Menands' Hill. They set up a midway, erected a fun house, the side show, the thrill rides, the coin tosses, cotton candy stands, and rides that towered over our heads, each tethered to a snorting diesel generator with some wild kid at the controls. It was heaven to a ten-year-old with ten dollars to spend. It was the yogi that I will never forget. With a blowtorch, he heated iron bars red hot and stepped on them; he blowtorched his own moustache and nothing singed. He stood on red-hot swords. The whites of his eyes were yellow. Too much heat, I figured. My mother stayed after the show; she wanted to know how he did it. The yogi stepped forward; we could see he was weary. No, there was no trick; it was the result of a great deal of training. Here, he was just being paid to do it. "Of course," he said, "It will do me no good, the money. I have used my gifts for financial gain; this should never be done. There is no hope for me."

I looked into his tired eyes. They were like black marbles: shiny, lifeless, and cold. A sudden chill gripped my mind; this man was telling me a truth. Special gifts are not given or gained to be used for popular entertainment; money made in this way is worse than no money at all. I had met my first Eastern adept, and we communicated just fine. He was working in the sideshow, still faithful to a system which could both empower and undo. I was still in cotton candy land, but I knew that he knew something I dearly wanted to know.

Twenty-five years and many lifetimes later it came back to me as I was working on the original manuscript of this book. Once a prominent drug educator, I found that my interest in the way that molecules could alter perception was leading me deeply into neuroscience and the biochemistry of consciousness. When a small group of frustrated editors, designers and helpers appeared one day, we gave them some space and *New Age* magazine came together in our offices. Released from the macrobiotic regimens of the *East-West Journal*, they rapidly expanded their scope and I was suddenly meeting and greeting all manner of spiritual traditions on an almost daily basis. My own ancestors included some prominent theologians, and I welcomed learning from roshis and rishis as well as numerous *New Age* crossovers, observing their differences and commonalities. I once nearly pranked editor Eric Utne with an article about a new group I claimed to be documenting before admitting that "*The Way of the Reed*" was imaginary, a mix of Timothy Leary, Carlos Castaneda, and Werner Erhard's EST seminars logo.

At the time, and even now, despite growing interest in altered states of consciousness from religious visions to yoga and meditation, few traditional religious scholars were doing any substantial work in the mind sciences. At the same time, most scientifically trained authors promoted explanations narrowly framed within their own religious traditions. *Why God Won't Go Away*, for instance, presents a radiologist's interpretation of brain activity as possible evidence of God; of little help to Hindus or Buddhists who don't believe in a single God, or God at all.

To balance my curiosity, I realized I needed a better grounding in religious belief and practice. I entered the Harvard Divinity School in 1976 and took courses with prominent scholars George MacRae, Harvey Cox, and Richard Reinhold Niebuhr. In 1981, I met the Dalai Lama and founded a small Buddhist institute to allow a scholarly lama to teach in the Boston area. His Holiness has an eager curiosity, and his interest in Western science led him to review parts of this work as it was being prepared. The Institute provided further opportunities to meet and speak with prominent scholars and practitioners from Eastern meditative traditions. I was able compare their methods and theories with typical Western practices, from prayer to ecstatic speech and dance. These observations and experiences from many perspectives revealed striking similarities just as emerging studies in computer-aided neuroscience were starting to reveal the basic structural connections and capabilities in the physical brain. It was finally making sense.

Our consciousness is always the result of the brain that makes it. There must have been some upgrade that allowed us to ponder the questions that our religions answer, and it seemed to have appeared about 100,000 years ago. If what seemed to be the case were true, it could also pinpoint how we, as a species, gradually acquired the concept of chronological time itself. If we could create "time" within our consciousness, this alone could generate valid insights into some of the major metaphysical rules underlying the world religions. Could we be eternal? As is so often the case, the breakthrough was accidental.

Instances of spiritual insight are often inadvertent, occurring during times of extreme emotion, stress or isolation that could create neurological distortions. Ethyl chloride is a simple medical spray that evaporates quickly to create surface numbness. A physician who used it commented that when inhaled, it produced the illusion of the mind operating at a distance. It was the 70's, I was a drug educator, I was healthy, and although wary of chlorinated hydrocarbons, I reasoned if I sat with my back to the wall I couldn't fall over. The first time was interesting, but the second time I completely overdid it. The room dissolved around me and I found myself in the sweetest everlasting floating forever-ness I couldn't imagine. I remembered this present world and life with nostalgia, but I was finally fulfilled, returning to my long forgotten home, and love was all around me. This went on a very long time of course until, with a sort of gentle tinkling sound, the world re-assembled like a mosaic coming together. Part of me was filled with joy and deep relief to discover there was a well of everlasting love inside us. We would all go home. The other part was flabbergasted. I'd had a genuine self-induced near-death experience sitting in my living room. I'd unplugged the top of my brain and I was hurtling past "near" to "real death experience" but, like a jet aircraft executing a fast parabola to create a few moments of weightlessness, I'd leveled off and come back to life. Could this be the perfect fulfillment we all desire, the universal timelessness of a simpler mind?

I was nearly certain now what actually happened in death, but was this a gift, a reward, or a curse? I wished my father were around to ask, but he had died when I was only twenty-two. After his death, my mother lived another twenty-six years. She was there to the end as McKinney & Son went bankrupt and was sold for the price of a parking lot. She had taught natural childbirth in the forties, natural foods in the fifties, and campaigned against additives in sixties; always ahead of her time enough to be a amateur savant without the degrees or time to stay focused on anything long enough to achieve professional respect. She was self-taught in medical matters, with several thousand dollars' worth of medical textbooks she filled with underlines, highlights and margin notes. One article in *The Lady's Home Journal*, describing the hormonal and physical effects of unintended pregnancy, sounded a note of caution to single women. Author Helen Gurley Brown was so irritated, as she recounted in *When Everything Changed*, Gail Collins' 2009 history of women's liberation, she decided to write *Sex and the Single Girl*. Mom never knew she'd been a catalyst for the women's movement. Her last preoccupation was her eventual stroke, a subject which kept her both stressed and stressful. At seventy-five, she agreed to try some meditative techniques I had learned from the Dalai Lama based on focused mental imagery. It worked, she said, and claimed her Holter Blood Pressure Monitor even recorded it. A year later, she was gone.

Immediately after the stroke, I reviewed the results of her CAT scans and was appalled at the devastation. Fully a third of her right hemisphere was gone for good. The attending neurologist said to expect the worst; no emotional affect and a foggy mind at best. The best thing, he said, would be another stroke. She was still having difficulty opening her eyes, and one side of her body was limp as a rag; she was sometimes speaking in French. By the third day, she was coming back. I bent over her when she seemed lucid. "I checked your scans, Mom. You've lost a big chunk in the middle of the right hemisphere, but your prefrontal lobes are fine and the visual cortex is still there." With her eyes still closed, she whispered "middle cerebral artery." She was right, of course. Then she asked, "Should I do my *vipassana* now?" I was floored. I had taught her to recall an image from memory and study it in the mind with the eyes closed. Even with such destruction, the teaching was intact, and so was her logic. I gave her hand a squeeze. "Wonderful, mom. It's great exercise for the visual cortex. That's just what you need now." Facing ahead, her eyes still closed, she said gravely, "What I need now are prayers."

She had read earlier drafts of this book, and she knew that we had what seemed to be scientific answers for a number of very basic human questions. She had taken the original chapter on death to the dying and had told me of their tears, sometimes of relief when someone saw that the end, when it came, was nearly a guaranteed heaven no matter what. Even those who had led a lifetime of faith were comforted; it provided a little insight that made a mystery less mysterious. My mother was religious, but for her the theories made sense; and she shared them with those whom she knew needed some universal comfort that might appeal to a curious mind. Now, in the anticipation of her own death, my mother was slowly returning to the faith she had always known.

She did not die of another stroke. She died a month later from bacterial and fungal infections that had been diagnosed but not adequately treated. It was as gentle a death as one could imagine as the pathogens slowly turned her brain to Cool Whip one cc at a time. At the very end, the last day I knew she was there, she looked vacantly into my eyes. I looked deeply into hers, and there she was, like a person at the very bottom of a swimming pool. She was looking up, letting me know she was there but, honestly, very far away. It carried another message. "You were right; I'm in another place." Late that evening, I could feel her soul sighing into the night with the sounds of the late night traffic on the long bridge in the distance.

The next day she was flatlined. Her pacemaker had Energizer Bunny batteries, however, and so she stuck around for curtain calls. She was an actress, and she had the whole stage to herself. During that last week, she showed up in four different people's dreams. "She said she was satisfied with her life and generally pleased with the way her sons were getting along ," said Prabha, an Indian neurologist who had become a close friend and confidant during her last four years. "She said that there was one small disappointment, however; she was sorry that your book wasn't published." She'd known I was trying to cheer her up but she hadn't let on, an actress to the end. The next month, the new owner finally gave up trying to stamp out pennies at the steel company he had bought for a dime, and the doors at James McKinney & Son closed forever.

The book has been published now, or you wouldn't be reading

Prologue

it. Like my father, I wanted to answer a question for all the people; and by the time my mother died, the answers were in hand. At the end, she was comforted in her sincere Christian faith and went, as her friend Judy said, "to the arms of the Savior she knew and loved." I may well do the same; those are my earliest memories at Sunday school, long before I was interested in girls or metaphysics, and I'm not going to try to modify them. I know where I'm going, and whether it's called endless lifetimes or the life everlasting, it's not a bad trip at all. The big question was answered as far as I was concerned, "and the rest", as Rabbi Hillel remarked "is only commentary." In fact, the process of answering one difficult question required answering quite a few more questions along the way.

"Is it possible to come up with a comprehensive philosophy of life and mind and fit the basic theory into six pages?" I had found a scholar who could tell me the truth. It was 1981 and I was still nervous about watching it come together. "Sure," she replied, "but you might need six hundred to explain just how you got there." It seems if we can agree to accept the concept of a virtual reality as the basis for our conscious experience of life, we are also witnessing the unexpected appearance of a novel systematic philosophy based on neuroscience. That begins to explain these ten chapters and why *You're Going to Heaven Whether You Like it Or Not* is about so much more than just the death experience. As to why it took forty years or so, it had to be rewritten and revised a few times. Science tends to advance, and this revision of the originally published version is completely up to date. I owe my sincere thanks to numerous recent readers whose enthusiastic response to a book over twenty years old inspired this updated and expanded effort.

The manuscript of the first edition was almost ready when, one day, I noticed the legendary psychologist B.F. Skinner walking through Harvard Square. I knew that he was not well and it might be my last chance to ask a question. "Dr. Skinner," I started, "I was an English major, like yourself, who got caught up in neurological detail. You once considered writing as a career. What effect did this have on your later work?" He smiled, and there was a real twinkle in his eye. "I have lived a long and predominantly rewarding life," he replied, his words flowing in precise intonation, "And I have always taken it for granted that a large measure of my success was due simply to the fact that I could write a great deal better than most of my colleagues." I shared a big grin with him. If it was worth the writing — if we were entrusted with that gift — the art meant as much as the science. If I could write, it was important to choose something important to write about; the more challenging the better.

He died a few months later and in honor of the craft of writing, I wrote the whole thing over again just to polish it up. If I'd spent half a lifetime answering one question, there's no reason not to be elegant about it and put on the best show possible. This book is easy to read, yet it may resonate deeply as it has already with readers all over the world. It took every bit of my skill, and nothing will ever be that hard to do or so rewarding to see completed. These essays, individually and as whole, will encourage you think about things you never thought about before in ways you never thought you would think about them. That is my first promise. The second is that if anyone worries that this synthesis attempts to replace faith with science, they needn't fear. There is nothing new here; all real truths are ancient. We simply change the explanations so that we can believe a little better whenever it's important to have a reason to believe. In these times, it's more important than ever.

The night I met the yogi, my mother and I rode the Ferris wheel up into the night, and at the top, it stopped with a shudder to take on a passenger. Rocking quietly in the dark, we looked down at the entire carnival, sparkling and bustling, the games, the tents, the support trucks and supply vans, and behind them the fields, the highway beyond, the Menands hill, and the starry sky reaching over our heads. It was very big and vast; and then, suddenly, the diesel gives a snort, the cab jerks forward, and we descend to cotton candy land again. This is not a long book; but for some it will provide a new perspective, a Ferris wheel for the mind. At least that is my hope; and then back to the lights, the action, and all the games of life.

Part One:
Introduction

1

Preparing for the Coming Faithquake

Faith Tectonics and New World Religions

> *"If God had wanted it that way, He would have made you all of one religion, but He has done otherwise so to test you in the various ways He has given you. Therefore, press forward in good works; unto God shall you return, and He will tell you about those areas in which you disagree."*
> — *The Koran, Verse (5:48)*

A pressing problem is facing mankind. It is a result of the powerful effects of a growing global society on religions originating in different parts of the world. Like the earthquakes resulting from periodic friction between massive continental plates, violent upheavals resulting from sectarian strife between religious groups continue to increase throughout large sectors of society. It would be a blessing if the next fifty years could witness the emergence of some advances in the effort to harmonize global faiths. Are we ready for this next step in our social evolution? Opinions differ, but the facts remain. Unless some substantial progress is made, local religious traditions could be the last casualty of the twenty-first century, surviving in the form of nostalgic cultural archives, greatly reduced in authority and influence.

It hardly seemed that way a few years ago. In a 1991 survey, 40 percent of those responding to a United States poll chose faith in God as "top priority." Twenty-nine percent chose good health, and only 2 percent chose money, prompting Wade Clark Roof, Professor of Religion and Society at UCLA, to write that such "astounding" numbers suggested a "cultural shift." Eighty-six percent of Americans responding to a survey commissioned by The City University of New York declared themselves Christian; but there are over three million American Muslims, and a million Buddhists as well as a half million Hindus. By the end of the 1990s, Christian fundamentalists in Vista, California, had elected enough candidates to a local school board to require the teaching of Bible-based history, while a rabbi in Israel condemned a dairy in Jerusalem for suggesting dinosaurs on its milk cartons were millions of years old, "despite the fact the world was created only 5,753 years ago." Meanwhile, Egyptian fundamentalists in Abu Zabaal prison were recruiting supporters when a thug cursed Islam. The resulting brawl, which lasted three hours, left three dead and eighty-five wounded. Religious belief appeared to be on the rise everywhere.

Yet by 2017, with the entire world on alert over increasing terrorism associated with the rise of ISIS and the phenomenon of global jihad, 72 percent of Americans responding to a recent Pew poll felt religion was losing its influence. Those marking "none" when ask to specify their faith had become, after Evangelicals and lapsed Catholics, the nation's most common religious category. Harvard Divinity School marked its 200th anniversary in 2016-17 by accentuating pluralism and tolerance for all, even considering a suggestion that the term "divinity" itself might be outdated for a scholarly institution. Their most recent student group, called 'Nones,' for choosing "none" as a religious preference, was growing in number. The trend is not limited to the United States. Forty percent of Britons reported they had no religion at all. If things are getting so relaxed, why are religious fundamentalists getting so aggressive? It is because they feel threatened, and they have good reason to feel this way.

Even as the faithful gather together worldwide, the orderly integration of world religions, once separated by sheer physical distance, has gotten completely out of control. There are more Muslims than Episcopalians in the United States, and Mormons are expanding rapidly in Brazil. Evangelical Christianity imported from America enjoys phenomenal growth in Korea, while Korean Sun Myung Moon, in America, preached a married Christ who was himself. Six million Lutherans pray daily in Indonesia; Japanese Buddhists teach Angelenos to chant the Lotus Sutra, and German Neo-Hindus in saffron jhabalas chant Hare Krishna in Red Square. Meanwhile, a new cold war is springing up between newly re-established Russian Orthodox priests and popular tele-preachers bent on purloining their new-found flocks with heavy-metal hallelujah revivals.

It's open season for souls, it seems, but nobody has taken account of where this is all going to end up. If human society does not come up with some generally acceptable understandings and accommodations between major world faiths, millions of people will continue to suffer needlessly and perish in confrontations based on ancient religious disagreements. Sooner or later, we will have to stop debating whose God is God and which Holy Scriptures are the ones to trust. Theologians and philosophers are beginning to perspire noticeably as we near an inevitable spiritual showdown.

The Roots of Regional Religious Tradition

Prominent Harvard scholar E.O. Wilson thinks a series of climatic changes may have sparked the rise of Western religions. At the Harvard Divinity School's 175th anniversary, he described the desert-like Middle Eastern biblical lands as lushly vegetated in recent prehistory, a true Garden of Eden. Rapid desertification of the region acted as a psychological shock wave, dislocating cultures and formalizing religion when oral traditions still spoke of a time when life was very different. It was also during this period that the Mediterranean cut through the Bosphorus to an inland lake in a massive torrent that created the Black Sea. This flood, described in both Biblical and Persian traditions, drove the rapid and necessary decampment of thousands of settlements with whatever they could take with them, the true Noah's ark. Recent DNA research suggests a migration of domesticated animals from that region to Europe at about the same time.

Likewise, later cultural upheavals resulting from invasions of Aryans, Mongols, and Muslims had a similar effect on religious orientation in Asia and South Asia. In historic and even contemporary times, events leave whole societies hungering for spiritual guidance to help counteract very real fears. According to religious scholar Alexander Berzin, archivist to the Dalai Lama, the major Hindu and Buddhist esoteric and tantric traditions appeared in the eighth and ninth centuries when it was important to find ways to unite differing peoples and castes to face impending dangers from invasions. When God only knows what's going to happen next, it's important to learn new ways to call on higher powers for help.

As a result, all major world religions originate in specific geographic locations with unique cultural traditions built in from the beginning. The Mediterranean basin was the cradle for early Christianity, which followed Greek communities to Rome, the rest of Europe, and ultimately to the Americas. From the Arabian desert the Prophet Muhammed's message spread south and east; from Arabia to Persia, India, Indonesia, Malaysia and the Philippines. The teachings of the Buddha traveled trade routes south to Sri Lanka and Thailand and the Silk Road east to Mongolia, China, and Japan. Lao-tzu and Confucius were born in China, and their wisdom went West against the flow of Hindus, Muslims, Buddhists, and Christians as well as Jews displaced from Israel in Eastern Europe.

The slow spread of regional faiths on foot and by primitive forms of transportation allowed large areas to become associated with one belief or another, a geographical religious homogeneity which has lasted until the present. Culture, society, and faith were all connected; one was a believer or not and there were no alternatives. When differing faiths encountered each other, there was usually either conversion or persecution. Western faiths especially have problems with synthesis; fundamentalist Muslims vie with fundamentalist Christians in bitter us-versus-them theologies even among themselves. In the East, mergers were often tolerated, although sometimes uncomfortable. In the primal case of the uninvited dinner guest, the unruly South Indian Dravidian deity Shiva, faced with the gradual encroachment of the grand new gods of the conquering Aryans, simply moved in with Brahma and Vishnu - trident, tantras, ashes, and all.

As Christianity worked its way into Europe, Celtic pagan religious feasts were recast as Christian holidays while sacred grottos such as Lourdes, once dedicated to local female deities, became associated with the Virgin Mary. In Tibet, animist mountain demons in the Himalayas were converted by Buddhist yogis into heroic dharmapalas, guardians of the Dharma. Only rarely did the encroaching religion totally usurp earlier beliefs; more often it absorbed them after converting the ruling classes.

Nepal, for instance, was Buddhist until King Jayasthiti Malla decided in the fifteenth century it would be nicer to be Hindu. Traditional Nepali Buddhists stayed put, since the Hindu gods were always part of Nepali Buddhism - but unable to compete in Hindu caste ranking, they have suffered socially to this day. Mongolia became Buddhist when a Tibetan adept won a religious contest;

the gentle Dalai Lama was suddenly Pope to a population of rambunctious Mongols. Perhaps the smoothest mass conversion of all occurred in the year 1000 when the Icelandic *althing*, their parliament, met and simply voted in Christianity for the entire country. In one rare instance of East-West accommodation, the message of Christ may have transformed a romantic North Indian shepherd deity into the divine Krishna. Indian historians have determined that the earliest records of this long established cult began surprisingly close in time to India's legendary contact with the Christian missionary-apostle Thomas about 52 AD.

This might help explain why Krishna's compassionate counsel to Arjuna in the Bhagavad Gita seems at times a South Asian Sermon on the Mount inserted into the epic war drama of the Mahabharata, otherwise devoted to impersonal concepts such as dharma and karma. India may have accepted the message but not the messenger. Thomas himself, according to history, established churches from Kerala to Chennai, where he died in 72 AD. Was martyrdom his glory or his karma? Nobody knows. Kwan-Yin, the Chinese goddess of mercy, is occasionally found depicted with a child, an iconographic synthesis based on Christian madonnas introduced along the Chinese coast by Portuguese traders in the sixteenth century.

When the East rejected a religion, it was often in reaction to a foreign culture rather than a foreign theology. "First come the priests," warned the first King of Nepal, Prithvi Narayan Shah, "then come the cannons." When the shoguns of sixteenth century Japan shut out Christianity, it was part of a total exclusionary policy so complete that Western technology was banned at the same time. Commodore Perry discovered Japanese samurai still hacking away at each other with swords in the mid-nineteenth century, hundreds of years after the development of reliable firearms.

As a result of this natural tendency to ground in a particular geographic area, each religion in the world today is expressed and experienced through the deepest traditions of a specific regional culture. As Internet and media merge us into an inevitable world consciousness, these regional beliefs may become our last links with our sense of individual identity. Given the unsteady world conditions prevailing, it should not seem surprising that more people than ever before have discovered both the cultural security and, for many, culturally appropriate answers available through religious belief and practice.

There are two sides to this harmonious picture, however. In

one sense we are all cheered to see the Russian Patriarch again leading his flock in Moscow, the Dalai Lama meeting with leaders in religion and science, or Mother Teresa's epiphany in the slums of Calcutta. Yet in another, slightly more sinister sense, this may also represent a sort of spiritual time bomb. There's room for only so many heavens on one earth.

Can We Talk?

It would be interesting to invite Jesus Christ, Muhammed, the Buddha, Lao-tzu, Confucius, Moses, and Manu for dinner to see whether they might come up with something like United Religions. Our dinner guests would probably think it was a fine idea, but since each is a representative of a higher power, they would have to report back to God, Allah, Tao, or Dharma for the go-ahead. Things might get stuck at the metaphysical level. There is good reason for this.

Western religions rely on mutually exclusive personal revelation to holy individuals such as Moses, Jesus, Muhammed, or Joseph Smith from one all-encompassing God. They have also tended to build on each other. Christianity added Jesus to Moses; Islam added Muhammed to Moses and Jesus. In America, the Mormons added Joseph Smith, and Christian Scientists in Boston cheered for Mary Baker Eddy. The late Sun Myung Moon said he actually *was* Jesus, although followers of the late Texas Branch Davidian David Koresh disagreed with the Korean leader. Neither has risen. Some believe that God still speaks to selected Westerners, including the top Mormon, but most claiming conversation with a deity these days are offered Prozac® more often than prayer.

Finding agreement among Asian believers is no easier. Most Eastern traditions replace mutually exclusive prophets of God with mutually exclusive interpretations of Dharma or Tao, the eternal universal system uniting the human physical and metaphysical experience. Hinduism is technically *Sanathan Dharma*, or the "traditional system." Buddha preached the Buddha Dharma, his own understanding of the way things worked. The parts are not really interchangeable. Theory and practices differ. Brahman is not Nirvana, the Tao is its own way, and none of them interface with each other.

Getting the original sources to cooperate could be even harder.

Yahweh and Allah might agree to the same menu, since neither like pork, but Ram might have a beef with the burgers because Hindus don't eat cows. Getting served could be dicey protocol, since God wants no other gods served before Him, but it makes things simple for the Buddha since his monks' rules say to eat anything put in the begging bowl, as long as it's in the morning. They don't eat anything at all in the afternoon.

Still, once the deities worked out the seating, they would soon discover how similar their messages are at the human level, the only level we humans can be concerned about. Like the larger sects of a major religion, the religions of the world today all seem to lead us in the same basic direction but continue to disagree on who is to be our guide and which guidebook we are to use. The more we investigate the basic dogmas of the world religions, the more depressing it becomes. Each originates in a different land, embodies idiosyncratic traditions, and is, to the devout, the only one there is. Furthermore, there hasn't been a really new world religion since the Sikh Dharma, Guru Nanak's alloy of Muslim and Hindu faiths. The Bahais have tried very hard but, as with Esperanto, attempts at cultural amalgamation this far along lack a certain spark. In fact, nearly all recent religious innovations have been no more than new interpretations of already extant theologies. Mormons, Pentecostals, and Christian Scientists all worship the same Jesus as Roman Catholics, and yet none of these sects existed two hundred years ago. The ecstatic devotion of Lord Chaitanya for Krishna in the fifteenth century gave birth to his egalitarian Hindu sect, while the pantheism of Ramakrishna established a national following in Victorian Calcutta. In nineteenth century Germany, Reform Judaism was born. In the twentieth century Japan, a Buddhist sect, Nichiren Shoshu, acquired a lay auxiliary, Sokka Gakkai, expanded, evangelized, and practiced politics until scandals separated the priesthood from the promoters. China, home of two ancient traditions, has recently given birth to Falun Gong, an entirely twentieth century synthesis.

There has been nothing radically new on the horizon for quite a long time and the pressure is building. In fact, the world might be due for a major religious event of some sort. In regular oscillations, periods of human pride in technology and power seem to alternate with periods of religious resurgence. The Renaissance gave impetus to the Protestants just as the industrial revolution underwrote the populist Baptists, Wesleyans, Mormons, and Methodists. Whenever it seems that mankind is becoming too arrogant with material powers, there seems to be a social migration back to religious faith, often resulting in entirely new sects.

With the proliferation of so many different religions around the globe, one can only wonder if we may be closer to new fusions than we expected. Just as plate tectonics describes the intense geological pressures where continental plates meet, leading to massive earthquakes and tsunamis, world religious pressure seems to be building to new intensities. The "faith tectonics" of regional religions are already causing cataclysmic social eruptions, literally creating "faithquakes" all along the fault lines.

A growing worldwide shift back to stronger religious belief, especially as a force to promote national unity, might have been fine fifty years ago but the time has long passed for the promotion of a God who loves anyone especially. We are facing more than a cultural shift; we may be facing a basic paradigm shift as dramatic as the notion that unbelievers could also be saints.

The Medium and the Meaning

Nothing happens in a cultural vacuum anymore. International marketing of everything from smart phones to pharmaceuticals fuels the engines of a dynamic world culture; video and Internet expose us all to each other all the time these days. It is not the word of anyone's God which motivates the millions, but mass promotion of images intended ultimately to sell products or politics. As advertising and personal ambition drive the world media, commerce overcomes creed in the multinational marketplace. Local heroes are replaced by international personalities; these days nearly everyone knows Madonna is a singer with a daughter named Lourdes, the Saints are a pro football team, and a Hail Mary is a pass to the end zone.

One unfortunate result seems to be a growing anxiety as to the source of any real truth. As psychologist Carol Moog points out, "The closer advertisers get to creating images of reality that coincide with people's perception of what reality looks like, the harder it is for consumers to test the reality of the message and dismiss it as advertising." Continually faced with censorship of information for political purposes or manipulation of information for commercial purposes, there is a growing international anxiety as to what is really going on anywhere. Laced by the Internet and examined every second by scores of satellites, the world is more than ever able to reveal the truth, and yet at the same time we seem faced with a different crisis of faith - faith that there are any real, believable answers anymore.

One overt reaction has been a growing distrust of many aspects of global culture. Nationalism is on the rise in nearly every region and it is not simply brexit or a wall along the border. Racial, religious, and cultural violence escalate yearly as people react violently to any suggestion of broadly shared goals and common needs. Arab nations withdraw into differing interpretations of the Koran to justify horrific conflicts, religious blood feuds fuel clashes in Syria, Iraq, Iran, Yemen, Afghanistan and Thailand. Survivors of Balkan and Chechen atrocities join forces with Al Qaeda, Isis and Boko Haram. Tribal violence tears at Africa year after year while religious outbreaks erupt in India as regularly as the monsoons. Europeans, once welcoming, turn edgy if not xenophobic, fanned by a refugee crisis of staggering proportions made up foreigners and their families fleeing for their lives. Between one quarter and one half of Austrians questioned in a poll said they would not want to live next to Turks, Poles, Romanians, or Yugoslavs. In the United Kingdom, fear of immigrants led to the unprecedented referendum that severed Britain from the European Union, stoking divisions with Scotland, Ireland and Wales.

Meanwhile, in the United States, immigrants are demonized by rightist Internet pundits, and laws have been passed establishing English as the "national language" just when more languages than ever before are being spoken among ethnic groups. In a survey of 1,500 Americans by the American Jewish Committee rating attitudes towards other cultures, 40 percent expressed negative feelings towards "Wisians," a non-existent group added to the survey as a response gauge. Having discovered more countries and cultures among us than ever before, the concept of being adult – and responsible to forces higher than self-interest – becomes increasingly unenforceable now that no one really believes that Allah is going to punish the Jews or that Jesus saves only white Republicans. In effect, the ultimate parents have checked out, leaving us all searching for rules that we once knew by heart, and, more to the point, the mechanism of threat or reward to uphold them.

The most unsettling by-product of growing global religious pluralism has been the erosion of a believable cause-and-effect system for ethics, morality, and social behavior. This in turn has led to a rather disturbing aspect of our current world order. Nearly everywhere we find the codifying of local culture into ever more conservative legislation, promoted by brave new alliances between religious fundamentalists and nationalistic isolationists that lead inevitably to repressive national policies. The pervasive use of local, religion-based cultural "morality" as a broad excuse for greater governmental control over aspects of personal life is being promoted globally. Gays are threatened with death in Uganda daily, and blasphemy prosecutions are on the rise in Pakistan. In the West, media saturation fills the gaps with crime and violence, making gun-slinging police the enforcer gods of life and death.

In one nation after another, this regressive trend is fervently supported by local firebrands warning that traditional morality is being threatened in a world that has lost direction and seems about to fall into a hedonistic, techno-humanitarian, sinful, pluralistic chaos. In response, we try to make dissonance and dissidents illegal and let loose the inquisition of the state against the misbehavers. Since God won't strike them down, we have to take on the job ourselves to shore up cultural habits and national mythologies with new laws, police, and prisons.

Commenting on these trends, Gary Bauer, activist President of the conservative American Values lobby, described the scene in the United States in the nineties as "an ongoing cultural civil war revolving around morality," pitting Americans "with a fairly traditional religious faith who feel that culture is out of control, against secular people who believe in a pluralistic society." In response to concerns that a free society was not going to welcome the idea of more legislated morality, Bauer was unconvinced. "We're pretty far away from worrying about society erring on the side of too many restrictions. The pendulum is not moving that way. Popular culture is still in the other camp."

His assessment underscored a growing association of religion and state policy as faith and patriotism become ever more dangerously mixed. While American evangelicals root for Israel, one Middle East nation after another is torn by insurgents demanding Sharia Islamic law; in Bangladesh, agnosticism can be apostasy; blogging the wrong views invites assassination. In China, where non-religion is the religion, the Central Committee of the Communist Party issued a twelve page directive ordering authorities to "resolutely attack counter-revolutionaries who make use of religion." When not terrorizing followers of the neo-Tao Falun Gong, it was reprimanding schools for allowing the recitation of religious texts on the basis that such practice was dangerous to mental health. Meanwhile, Muslim Uyghurs started attacking Chinese atheists with knives near the borders of Buddhist Tibet.

In our global society, the very regionalism of the great religious faiths now provides the major irritant of social misunderstanding, too often promoting national homeboy fantasies of a national purpose under a national God. Rejecting the notion that the earth is of common concern to us all, more than ever before we are urged to identify with simple, marketable cultural icons and ethnic symbols. For emotional outlet, national flags and colored ribbons replace personal tributes and statements.

The trend is not limited to adults. Rather than embrace the mixing of cultures on the modern university campus, students are creating ever more exclusive enclaves of their own, studying and socializing only with others like themselves. "It's just backwards," wrote educator Theodore Sizer. "We need more diversity... we need to train people to resist the powerful images commerce puts around us." Ernest L. Boyer, former United States Education Commissioner, agreed. "Separatism and even tribalism in the old-fashioned sense are increasing," he noted. "The implications are frightening. If humanism and communal understanding cannot happen on a college campus, how in the world can it happen on city streets?"

In fact, a variety of this chilling trend had already been institutionalized in the Arab world as thousands of ultra-conservative Wahhabi Muslim madrasas, religious schools supported by oil-rich Saudis, graduated the first wave of jihadi terrorists to wage global war against all infidels. Our sense of personal and social security, once synonymous with cultural identity and religious belief, faces ever greater challenges today in a world that is ironically both increasingly connected and increasingly alienated.

During the last decade, reactions from local cultures to foreign religion have ranged from terrorist violence to dismayed accommodation. Immigrant Muslims face skinhead lynch mobs in Germany while in Nepal, traditional Hindus endured temple "sit-ins" by immigrant Western Hare Krishnas demanding equal rights. The Saudi government allows only Muslims into Mecca; to get background shots for Spike Lee's film, *Malcolm X,* an entire film crew converted to Islam: mercurial Muslims at best, but Muslims nonetheless. Allahu Akbar!

Most of the time, it is more depressing. The Arab Spring spun out of control in Libya, plunging it into regional chaos, and a military dictator seized control in Egypt to block Muslim fundamentalists. In France, politician Marine Le Pen resurrected her father's race-baiting invective aimed mainly at Islamic immigrants, while in the 2016 US presidential elections, two out of three leading Republican candidates - Donald Trump and Ted Cruz - campaigned on platforms that were considered isolationist, nationalistic, and racist at least in appearance, if not in fact. Popular culture seems everywhere to be testing new levels of paranoid repression and cultural intolerance.

This leads to an interesting speculation. Could these pressures become the catalyst for new religious directions as well? Could there arise new and original forms of spiritual understanding and assurance that could be expressed and shared across cultures in an increasingly interconnected, increasingly intercultural, yet increasingly anxious world? As the world warms to the heartfelt compassion of Pope Francis in Rome, many hope this signals a trend; but to many others, it must seem that we are facing the apocalypse. In fact, we only need the determination and will to undertake a thoughtful re-examination of the universal human search for those eternal answers which only religions can provide. In doing so, we might finally learn to share some of the more reassuring, more universal, and above all more believable aspects of our faith, phrased in some universally expressed wisdom we could all embrace equally.

High Tech Hybrids: If Ever the Time Were Ripe

Social consciousness everywhere is trending to global concerns broadcast by the Internet and social media and directed toward medical emergencies, worldwide climate change, and international refugee aid. The increasing use of modern technology in our lives, across borders and across cultures, has already demonstrated how differing cultures can communicate on a platform that shares a common language. Likewise, any new expansion of religious philosophies will also likely embrace the sciences as a tool for discovery and compassion rather than simply the cutting edge for commerce or for war.

So far, at least in this century, the major role of technology in the service of religion has been to increase the reach of already established world faiths. There has been no United Nations "Religion Project" to coordinate the theologians, psychologists, philosophers, and religious leaders. IBM's Watson supercomputer hasn't precisely located the soul yet, while the only contender for a sect with a high-tech terminology would be L. Ron Hubbard's neo-rationalist Scientology. Unfortunately, despite the apparent utility of some of its simpler auto-hypnotic practices, it lacks a comprehensive philosophical and ethical structure and cannot, therefore, attract broad social support.

Still, as the twentieth century drew to a close, a number of synthetic "religions" had already appeared. Each had its own techno-jargon, and promoted itself with seminars based on original mind science philosophies from Werner Erhardt's EST to Neuro-Linguistic Programming, each claiming a "scientific" basis. In precisely the manner that medieval mountebanks adapted "hocus pocus" from the priest's *hoc est corpus Christi* - "this is the body of Christ", creating pseudo-Latin to mislead village oafs, scientific sounding philosophies are usually more convincing to seekers who are functionally illiterate in the very sciences cited to buttress their beliefs. By now, most of them have simply died out.

It is a shame that nothing resembling a global faith has sprung up anywhere recently, given the current circumstances. With satellite, Internet, smart phone, and social media, our global net is already beginning to shrink us into an interconnected people separated only incidentally by geography and culture. Differing traditions and practices are blending as never before. In Malaysia, for example, high school education was offered in Chinese, Malay, and English. The number of students in each category was roughly equal until parents began to discover that good English was worth more in the job market and more parents began sending their children to English schools. This in turn began breaking down traditional barriers between local ethnic Chinese minorities, whose children now shared English as a common tongue. All over the world it is happening.

It has happened before. The rapid spread of early Christianity was due largely to the existence of Greek communities in numerous cities lining the perimeter of the Mediterranean Sea. The Roman Empire, nearly at its historic height, interconnected the entire Greco-Roman world in a common law and language. In fact, it was two languages. The common language was Latin, but art, philosophy, and science were discussed by the educated in Greek. The message of Jesus, carried from city to city by the journeys of Paul and other early apostles, found a receptive audience mainly among Hellenic Romans and Greeks adrift without a religion that made any sense and too many philosophies that did.

If there were ever a time similar to the Roman Empire at the birth of the Christian era - the first time a faith could travel nearly everywhere in a short time - we have nearly identical conditions now with our massively interlinked global communications networks. In our current scenario, the Mediterranean becomes the globe as we face a scenario similar to the the plight of the early pre-Christians. They had all the Greek religions and the Roman religions and some Egyptian cults on the side. Those disliking the devotional ceremonies of the Mithraic mysteries often found doctrinaire Stoics a bit too Zen. Intellectuals rarely believed in Zeus, but criticized the Epicureans as "be here now" utopians of doubtful patriotism. To many Greeks, Judaism was appealing, but ritual circumcision was appalling. Still, many were attracted to its monotheism and sense of social justice. The messianic promise of Christianity, combined with the full richness of its monotheistic Jewish roots, was different and exciting.

Once Saint Paul pioneered baptism without circumcision, a crucial turning point in the faith, the Christian message spread rapidly around the Mediterranean from one Greek community to the next. Every book in the New Testament was written in Greek, the common scientific and philosophical language of the Roman Empire. In the first century, going Greek was going digital and the literati could read it anywhere. Ironically, Jesus, an Aramaic speaker familiar with Hebrew, could not have read his own Gospel except in translation. His message was far more relevant to a people he had never known than to his own Jewish co-religionists. The time was right, society was ripe for a change; and in less than a hundred years it had spread everywhere Latin or Greek was spoken. In three hundred years Christianity had become the religion of the Western world.

The world currently presents us with more than a dozen major world faiths, each with scores of legitimate variations, not to mention philosophical schools, cultural traditions, and regional cults led by local charismatics of every sort. There is no end to the choices available these days, from the God of Abraham to the Gods of Zoroaster. There is one vast difference, however, and it is in the power which organized religion actually holds in modern secular society. One of the more useful results of the intercultural blending among the nations of the world is an agreement on rule by law rather than by dictate. Since human law is traditionally enforced by secular authority, by the twentieth century traditional values were increasingly promoted by civil, rather than religious agencies, from international coalitions to confront disease and famine to local volunteer groups from Doctors Without Borders to the Girl Scouts.

Ironically, the most brutal behavior seems to originate with those promoting fundamentalist religious belief. The last twenty-five years have been, in this respect, rather grim. In that time we have watched Iraqi Sunnis kill Iraqi Shia, Israeli Jews bomb Palestinian Muslims, Rwandan Hutus slaughtering Rwandan Tutsis, Bangladeshi Muslims hacking Bangladeshi bloggers, Palestinian teens attacking Jewish shoppers, Buddhist Sri Lankans killing Hindu Tamils, and mad-dog Islamist ISIS fanatics murdering thousands in dozens of horrifying ways. Meanwhile, Christian Americans fight a war they execute by remote control, drone-targeting Afghani Wahabis, Yemeni radicals, and Syrian jihadis. Does anyone actually believe God, Allah, or Jehovah is behind all this? Not bloody likely, but try to convince a fundamentalist of any major faith that the heretical unbeliever may also go to heaven, and some holy quote will be produced proving otherwise. There is only so much flexibility available if one must ultimately rely on religious dogma.

Attempts at cross culture inclusion, even at the scholarly level, have often elicited criticism. In 1991, at the World Council of Churches meeting in Canberra, Australia, a Greek Orthodox prelate protested that Korean feminist theologian Chung Hyun-Kyung's depiction of the Chinese "bodhisattva of compassion", Kwan Yin, as an image of the Holy Spirit had gone too far. An invocation which included elements of Native American prayers to the forces of nature was similarly panned as nearly pagan. As our world culture grows, it is increasingly difficult to be a religious purist, and in response the purists become even more insistent upon getting back to fundamentals.

This is, in essence, the basis of the underlying problem. All attempts at world ecumenism are challenged from the start because they always start from the basis of one major world faith or another. A broad minded Buddhist cannot really be a Christian any more than a sincere Muslim could embrace Judaism. A religious person has to be a "this" or a "that". Less religious individuals have an even greater problem. To define oneself as agnostic or atheist seems to express an active nihilism that few actually feel. Indeed, many of those who are lukewarm about their faith would enjoy a deeper devotion, but do not know how to find it without the concern they may be abandoning their intellect in the passivity of dogma and ritual.

A Scientific Approach?

The major challenge facing any new perspective is that it cannot be in opposition to any current beliefs. It must come from an entirely new direction. It would be impossible for any new "religion" to supplant or absorb any major world faith because there are simply too many of the faithful. There could arise, however, higher-order philosophies based on generally accepted knowledge, knowledge that was not available in the past. Such revelation would not require a holy book of rules and religious history, nor a social philosophy set down by an anointed one and his followers in ancient times. If these premises were accepted as culturally transparent, they need not conflict with religious faith.

The most universally believable higher force for the past five centuries has been not religion but science, arguably the most powerful religion on the planet today. Imagine using scientific method to provide answers to major spiritual questions, to articulate a comprehensive philosophy of life, justify a moral code, and to provide inspirational practices for personal self improvement.

Much of the above has, in fact, been worked out in theory and sometimes in practice. What seems to have been lacking is the articulation of a believable structure that could tie together the nature of life, as we experience it, into some meaningful pattern using broadly accepted scientific principles. The Roman Catholic catechism asks "What is the purpose of man?", and Thomas Aquinas provides an answer. Until recently when it came to questions such as these, science drew a blank and religion stepped in. Many of the major conundrums that trouble us, we are told, are not scientifically answerable. These are, unfortunately, some of the real head-scratchers of human life.

What makes it even more difficult is that answers provided by religious texts tend to operate within a neighboring universe, a location replete with an eternal system that seems at odds with the laws of time and space. Science is here to define and manipulate temporal laws, but the holy books discuss truths and concepts such as the life, or the lives, everlasting. If science located an everlasting anything it would be difficult to describe. Worse, one could not publish the definitive study until after the end of forever, which sums up the situation. When compared against each other, science and religion often tend to make the other seem trivial in the most fundamental enterprises of human life.

Who has the answers? Does natural law exist by the will of God? Is there an ultimate reality that centers us universally and personally, that pertains to our enemy even as it does to us, that sets the limits of our existence in a manner we can intuit, for reasons which seem just? Is there a reason to be good? More to the point, is there any way we can find that point of reference that will allow us to answer these questions without offending either the rational or the faithful?

2

The Metaphysics of Neuroscience

Emerging Perspectives, Shifting Paradigms

> *"To study metaphysics as they have been studied appears to me to be like puzzling at astronomy without mechanics.... We must bring some <u>stable</u> foundation to argue from."*
> — *Charles Darwin*

    The world could clearly benefit from some form of common religious interface; some means to separate culture from content and make it possible to share a basis for personal faith that is universally acceptable. Each year we are moving closer to a common world news cycle, could we also be moving toward a globalization of religious thought? In a world of over six billion humans, we should have enough accumulated experience to derive some general guidelines for good human behavior that transcend local tradition and national politics. The real problem, in fact, has very little to do with this sort of wisdom. Nobody really disagrees about naughty and nice. It's not even a matter of cultures. Religion is much more than theology - the study of the nature of God. Every religion on earth today that enjoys credibility, cultural acceptance, and at least a half-million followers is defined by three general categories of thought and practice. The first could be termed "Cultural Ceremonies." The second would be "Applied Social Psychology," while the third encompasses "Metaphysics," the theology or philosophy behind it all.

    By far the greater part of most religious activity in any part of the world today is taken up by the first two categories. Our cultural calendars are dotted with regional, national, and even international observances of religious rites and holidays. Christmas is a world event celebrated in Mumbai, Bangkok, and Tokyo; Muslims shuttle to Mecca from Morocco, Marseilles, and Memphis, Tennessee. Everyone has New Year's parties, feasts, saints' days, and local celebrations. If it doesn't disrupt the local social fabric, nearly any form of personal religious observance is respected. Cultural politics may clash, as in Iraq, Ireland or India - but as individuals, we have no quarrel with another's yearly cycle of faith and celebration provided they stay within the cultural expectations of our region.

The second area of religious practice, "Applied Social Psychology," is even less problematic. This is because the great lawgivers and prophets gained their followings through their vision of the universalities of human social behavior, the ability to break it down into simple rules, and the charisma to convince others to use these rules as a basis for personal and social guidance. Any savior too specific for general acceptance ends up with a cult, not a cathedral. Mother Anne Lee's Shakers are gone. There are no Essenes in Judea, Kadam-pas in Tibet, and we could fit all remaining Swedenborgians into a large auditorium. Nearly twice as many met at Woodstock in 1969 as practiced Christian Science forty years later. In many divinity schools and seminaries, acceptance of multiple faiths and beliefs in a global setting has made the pastoral impetus toward social service to those in greatest need their primary focus. The time of fierce theological debate is over; it would be considered impolite, if not improper, for an Evangelical professor to suggest that her Buddhist colleague convert to Christianity.

Tolerance of another's religious customs becomes a necessary fact of life with so many traditions mingling in the crossroads of our growing global society. We can't convert them all, and those religions that get pushy about specifics will simply lose out. Most at risk are those requiring a physical hereditary link for membership. This trend is especially pernicious to religions that do not accept conversion. Orthodox Hindus and Jews alike watch their numbers shrink with each successive generation. Orthodox Parsees, who require both parents to be Parsees, are an endangered species. Descended from the original Zoroastrians, they represent the oldest continually practiced organized religion on earth. Less than a hundred thousand survive and there's nothing any non-Parsee can do about it. The Parsees are hard at work crafting a solution.

By contrast, the world's largest organized religion, the Roman Catholic Church, is fighting for its intellectual survival against ancient customs that proscribe interpretation by other than a celibate male priesthood. The sweeping exposure of sexual misconduct by members of the church, and the efforts made to disguise or deny it, became a worldwide scandal of unprecedented proportions. In the United States, "lapsed Catholics," as a group, are the single largest denomination, by numbers, in the entire country.

History demonstrates that only theologies based on a clear understanding of human nature - and expressed in a manner universal enough for translation and adaptation - can hope to survive more than a few generations. This is why the Hindu Shiv Sena movement in India is as doomed as the Christian Science Church; neither will survive the twenty first century. The former is too violent to be considered truly Hindu; the latter too closely identified with the teachings of Mary Baker Eddy, a Victorian charismatic. Rules for human life and living, at the heart of all world religions, must be broadly based. Reviewing the second category, the rules for human behavior, once again it seems nobody has any serious differences. Jesus preaches humility, Muhammed promotes generosity, St. Paul wants us to love, Moses and Buddha remind us not to hate or kill, the Tao keeps us steady, and Krishna asks us to open our hearts to devotion. All advise us to help the weak, support the poor, heal the sick, and be honest with each other. Their rationales differ, but the results are the same. Most Hindus in Calcutta revered Mother Teresa despite her proselytizing, and the Dalai Lama won the Nobel Peace Prize despite objections from the Chinese. Goodness is recognized everywhere, kindness is always welcome, and love embraced by all as a human response to the tragic beauty of life. Seen in this way, all religions mirror an inherent human wisdom universal in nature and specific in the telling. We have no unbelievers here, nor any reason to disagree.

Between the celebrations and proper behavior, two-thirds of our religious faith and practice are nearly transferable from one cultural currency to another. Despite differences between a Jew and a Muslim or between a Hindu and a Catholic, we seem to agree on almost all the day-to-day questions of real life and how we are to behave towards each other. We would enjoy most of each other's parties and ceremonies as well. So what remains to quibble about? The only area in which religions really differ is in the third category: metaphysics. Metaphysics is the philosophical study of the abstract ideas that lie behind an expressed faith. At the most basic level, all religions try to answer three basic questions. "Where do I come from?" Why am I here?" and "Where am I going?" Expanded into the world around us it becomes "Where did it all come from, why is it all here, and where does it all go?" Bertrand Russell, the celebrated mathematician and agnostic, made a list of five questions he claimed science was unable to answer: "Is there survival after death; does mind dominate matter or vice versa; is there a purpose to the universe; is there validity in the assumption of natural law; and what is the importance of life in the cosmic scheme?" Only God or Dharma, we are told, have clues to those mysteries, and only those made acceptable by ritual initiations, rites, or specialized education are entrusted with the interpretation.

Unfortunately, the answers all seem to differ. It's apparently not the eternal questions that keep us arguing as much as the variety we find in the answers. That this nearly philosophical aspect of religious practice was the basis of so much suffering in the twenty-first century will be a source of wonder in the twenty-second. Still, at this time, most religions continue to insist their answers are the one and only truth despite how unlikely they may seem to others. There may be another possibility shaping up, however, which seems to have passed unnoticed. Since all major religions are adaptable, they must be somewhat pliable. Nearly all religious dogma is based on interpretations of statements or writings general enough to transcend culture. New interpretations are never unthinkable. Returning to the "faithquake" metaphor, just as pressure at the earth's core can make solid rock flow like hot plastic, so contemporary social pressures could become intense enough to force new adaptations and interpretations of the most traditional and accepted mainline faiths. Pope Francis and the Dalai Lama have both broadened the boundaries of their traditional dogmas, while the deadly actions of fundamentalist Muslims have prompted broad condemnation from the rest of their faith and reassessment and reexamination of how this major belief system is interpreted and taught.

If we could find some basis for a shared metaphysics, we might learn to appreciate our global religions for what they are - poetic wisdom distilled from our cultural ancestors who needed neither MRIs nor scanning electron microscopes to perceive the underlying wisdom of human existence. Wisdom is found in the universals, not the details, and a universal human metaphysics would be possible only if the details had no cultural basis or bias at all. There seems only one way to do this.

The Metaphysics of Neuroscience

Scoping out new answers to unanswerable questions would seem beyond the ability of any one person or even a group of specialists. Recently, however, a growing amount of interest has become focused on the mind sciences, especially recent investigations into the phenomenon we call consciousness.

There is a good reason for this. No matter what faith we follow, we are all aware that there are millions of people who believe otherwise and seem not only to survive, but to prosper. The ability to accept that there are a lot of different religions without making further judgments as to which is right or wrong leads to a simple line of reasoning: since it appears that all human cultures have religions, might then the origin of human religions be found within the nature of human consciousness itself? MIT's linguistic pioneer Noam Chomsky long believed the basis of language lay in a brain structure common to all, but faced embarrassing controversy at the end of his career as advances in neuroscience failed to locate it. Religions, in contrast, embody the many differences specific to their many cultures, yet all attempt to answer the same questions.

If we observe groups of humans in different areas over time, they will always find religion of one sort or another, even if they may disagree in the details. If this is the case, it would strongly suggest that the basic drives that lead to religious faith may be internally generated. It could result from a species-wide need-to-know, externalized through varying but remarkably similar social structures, customs, and belief systems whenever a human culture has reached a certain level of development. This line of thinking could, in turn, lead one step further. Any philosophical drive common to normal human consciousness would have to proceed from some activity at a neurological level, something in the way the brain operates normally, affecting all human thought in a similar way, which is then expressed differently through different cultures.

This is probably the path to follow, although it may be distasteful for the uninitiated. The dualism of Descartes, suggesting a mechanistic brain that fabricates the mind, gives Hindus hives and Christians the creeps. Seeking clues to the spirit or the soul in a mass of wires and plumbing puts off both Baptists and Buddhists alike. Still, it remains the most likely direct route to the root of religious experience as we perceive it. The brain is, after all, the only organ capable of conscious perception. We can't do it with our toes or our tonsils. In fact, the primary importance of the brain in the perception of consciousness in all its forms has been well known since the ancients. However, many of the actual methods by which the brain accomplishes its task have been revealed only recently through rapidly advancing medical technology. The proliferation of linkages between brain science and computer science has nurtured a powerful alliance during the last sixty years. The architecture of the brain is finally being defined, and it is beginning to provide us with some first clues to the language of the mind itself. In fact, it becomes ever more likely that philosophers or theologians of the twenty-first century may eventually be required to show fluency in mind science, just as modern medical doctors must know their biochemistry. Things have changed that much.

As a natural result, we may be drawing closer to new forms of synthesis, entirely new insights which might finally help harmonize scientific method and religious belief. The correct term for this would be "neurophenomenology," literally "using the neurological sciences to determine the nature of reality," and this is, in fact, what seems to be emerging. "Neurotheology" has fewer syllables, says basically the same thing and links it specifically to religious philosophies. Just as Thomas Aquinas developed his philosophical system, Thomism, utilizing Aristotelian logic to order and anchor Christian theology, so modern religious thinkers are starting to warm to the new horizons opened through the study of neuroscience to provide intellectually universal and generally agreeable concepts on which to compare and explore their conclusions.

The fundamental reason to use brain science as the basis for a comprehensive systematic phenomenology is the simple argument that since we only experience what we perceive, we should first examine the structure and function of our major organ of perception. In learning more about the way we perceive reality itself, we may discover clues leading to simple and believable explanations of otherwise traditionally unexplainable mysteries. There are limits to our understanding, but this may be more because of the way the brain arranges consciousness than from any lack of enlightenment, devotion, or grace.

Virtual Religion?

The concept of consciousness as the ongoing result of a process is just one example of the sort of systematic viewpoint which seems to be rapidly overtaking contemporary thought in this area. From the ecology of the earth to the networking of databases, systematic perspectives seem to be emerging in many areas. Still, would we be willing to give up individual saviors and prophets for a better system, even if it preserved the beauty and the wisdom of our ancient religious heritage? We have accepted our beliefs as reality; could we reweave that fabric with a common thread and still be as sure?

Furthermore, this line of thinking creates an inescapable speculation. If the rules of consciousness turn out to be flexible under some circumstances, then, given the proper conditions, might not any of us perceive a timeless eternity or enjoy a truly transcendental experience? The philosophical implications of recent research along this line of thought, taken far enough, are already suggesting there may be answers to most of Russell's questions. In the process of normal brain death, for instance, we will all experience temporary states of consciousness in which the perception of both time and space are greatly altered.

The real problem is that if we use brain science to answer the question "What really happens when we die?" and if such a theory became generally accepted, every religion on earth would have to either deny it or demonstrate that their holy scriptures could include it by flexible interpretation or adaptation of traditional dogma. In fact, as some of these theories were being worked out, there was, for a time, some serious apprehension that a genuine breakthrough in this delicate area could create volatile, perhaps even violent, reactions among devout followers of one religion or another. Copernicus waited nearly until his death to publish, and author Salman Rushdie was in hiding for years after being condemned for his allusions to the prophet Muhammed in a popular novel.

There's no question that people get very emotional about their religious beliefs. Going to heaven without believing in Jesus is impossible for a Baptist, and any suggestion to the contrary is heretical. If eternity were found to be a state of mind experienced during brain death, any sinner might deny his Christ and theoretically, if not theologically, make it home free. Belief is belief; and if a concept appears valid enough that it becomes widely accepted, the believer is as assured of heaven by that means as by any other. Would neurological answers be inherent heresy, repugnant to the sincerely religious of every faith?

There are, it turns out, few scriptural bars to contemporary explanations so long as they do not fundamentally compromise accepted religious belief. There are limits of course. Finding the bones of Jesus Christ would be cataclysmic, proving His existence to the unbeliever while also denying His ascension, a basic tenet of Christian doctrine. Such a discovery would forever be contested and the discoverer marked for life as the source of a serious schism in the faith. One cannot deny basic dogma and hope to escape censure.

Fortunately, finding a scientific explanation for the experience of an eternal afterlife need not deny the experience itself. It should, after all, be within the power of God or Dharma to design us in such a way that we might transcend properly to our heavens without smoke and mirrors. In Varanasi, for instance, many Hindus believe that Bhairab, the lord of death, allows those fortunate enough to die within the sacred city to avoid the tedious rounds of rebirth by simply collapsing their future lives into one amazing instant so that they can "see Shiva" immediately. This endless-lifetime-express has never been fully investigated, but insights into how Bhairab might accomplish his feat would not invalidate the event. Describing the method need not reject the miraculous. It can even reinforce and revitalize the faithful to realize that neurological insight can be comforting, just as the beauty they appreciate in a sunset cannot in any way be diminished by a basic understanding of the biochemistry of its earthly perception.

Historically, religion is usually accommodating. Only a few preachers had serious problems with Charles Darwin's theories. A more typical Victorian clergyman, Charles Kingsley, read the recently published *Origin of Species* and wrote to the author, "I have gradually learnt to see that it is just as noble a conception of Deity, to believe He created primal forms capable of self development into all forms needful pro tempore and pro loco, as to believe that He required a fresh act of intervention to supply the lacunae which He Himself had made. I question whether the former be not the loftier thought."

Darwin's theory, like that of Copernicus, provided a better system to explain fundamental aspects of the world around us. In the process, it triggered a basic restructuring of scientific thought similar to that which followed the discoveries of Copernicus and Isaac Newton. These abrupt and radical changes in scientific perspective were first identified and described by philosopher Thomas Kuhn in *The Structure of Scientific Revolutions* as "paradigm shifts." They occur whenever a new perspective forces a wholesale restructuring of the predominant viewpoint, such as the shift from Ptolemaic astronomy to the new Copernican heliocentric system.

"The Copernican Revolution", as it is called, was a fundamental philosophical event. Once man was no longer the center of the universe, all sorts of other assumptions began to cave in. Religion had always maintained a connection with natural science, and by the sixteenth century Christian theology had embraced Ptolemaic astronomy. The concentric crystalline spheres of the Ptolemaic universe had seemed a bit lonely, and so early Christian writers had populated them with all manner of heavenly winged creatures. Having deeded the heavenly real estate to cherubim, seraphim, archangels, and so on, they found it embarrassing to evict them all. For a time it seemed easier to evict the Copernicans, but too many telescopes confirmed the results.

As paradigm shifts are not improvements in the old system but the unexpected introduction of a new one, they inevitably face opposition from the many institutions and individuals who are associated in one way or another with the status quo. There were many universities at the time of Copernicus, and professors of Ptolemaic astronomy were the only ones available. Some switched over easily; others were dragged kicking and fussing into the new era. Some never do switch. The great Victorian scientist and explorer Louis Agassiz discovered the ice age and collected many fossils, yet he never accepted Darwinian evolution. Some scientists still have problems with quantum physics and think their colleagues are "entangled" with impossible theories even as satellites use it for standard hand-held GPS applications.

Even when science has shifted, it can take centuries for cultural interpretation to change worldwide. By now nearly all educated scientists accept the evolution of species, but forty-seven percent of Americans responding to a recent Gallup poll still believed that God created man "pretty much in his present form at one time within the last 10,000 years." Basic breakthroughs in science and philosophy can take time to earn wide popular acceptance.

Werner Heisenberg meets William of Occam

Paradigm shifts characteristically exhibit two aspects. They seem nearly obvious once described, and yet they always require the best science of the time to provide the information which makes the new perspective possible. When the new insight finally occurs, it often happens so dramatically that it seems sudden and unexpected even to the discoverer, although it is nearly always the result of many years of effort.

"At first I was deeply alarmed," wrote Werner Heisenberg, describing his initial insight into quantum mechanics. "I had the feeling that, through the surface of atomic phenomena, I was looking at a strangely beautiful interior, and felt almost giddy at the thought that now I had to probe this wealth of mathematical structures nature had so generously spread before me. I was far too excited to sleep."

Others report the same experience: the alarming discovery of a new way of understanding some basic phenomena, profound in implication and yet so elegant in concept that it simply must be right. The new theory itself often proliferates so fast that it takes time for the proof to catch up to it. The Copernican system, as it was first published, was faulty. It required the combined work of Johannes Kepler, Tycho Brahe, and Isaac Newton to both prove and improve a system so obvious it was already widely accepted. Copernicus had utilized the best observations late renaissance technology had to offer. It was enough on which to launch a theory more elegant than the technology itself could adequately support.

It was likewise not discovered until the twentieth century that Newton had himself fudged some of the experiments he described in the proofs of his *Principia*. The structures he had uncovered were so elegant that he would not let the limitations of his own instruments, too crude to yield such accuracy, get in the way of his new discoveries. It made too much sense to be wrong, so he ran with it even when he knew he might never be able to prove it. "I am a physicist," Einstein would say, "I lack the mathematics to prove my theories." Eventually the mathematicians caught up.

The elegance of the ideas which re-order the thinking of an era always reflect an inherent simplicity. In this, they tend to conform to the example of "Occam's Razor", a philosophical observation by the fourteenth century English cleric William of Occam. His insight, proven again and again throughout the history of science, is expressed most simply by the phrase "nature abhors complexity". By cutting away any excess, despite the myriad possibilities and varieties of various solutions, the most likely are those that accomplish a task with the least effort. Given a number of possible explanations for any phenomena, the simplest is invariably correct. The theories of Nicholas Copernicus, Isaac Newton, Charles Darwin, Albert Einstein, Niels Bohr, Werner Heisenberg, and Stephen Hawking all explain a wider range of physical phenomena with a more compact system than had been previously available. Each of these new perspectives allowed new and unexpected observations to fit into a radically different, but inherently simple, structure.

It is the second aspect of a paradigm shift that may not be as immediately apparent to a purely philosophical investigation. It seems that theories that change the way we think are nearly all catalyzed by very specific advances in technology. Without the lenses of Hans Lippershey in 1608, Galileo would have had no reliable telescopes. Without the improvements of Newtonian physics, there could not have been a nineteenth century Michelson-Morely speed-of-light experiment to provide new questions that Einstein finally answered. Like relay runners passing the baton, finer science creates finer theory, which in turn creates even finer science. It was only a matter of time before the tools of brain science could offer new perspectives that could make a paradigm shift in religious scholarship possible, if not inevitable. New understandings are emerging, as radically different from the traditional world view as the solar system of Copernicus was from that of Ptolemy.

The Neurotheological Paradigm

Our sense of reality is generally accepted as our ongoing reaction to what is happening in the world around us. We are taught that the manner of this reaction determines our evaluation of a person's mental state. The world is real, but we interpret it differently; the universe is relatively fixed in time and space, and the way we interact with it is the variable.

This is the way current philosophy works. From a neurotheological viewpoint, this is backwards. The only place to start is to begin by acknowledging that the reality we perceive at any time is actually taking place in the brain. It is a virtual reality, perceived by a consciousness with rules and limits determined by what is available, neurologically speaking, to work with in any given brain at any given moment. Reality is not decreed, it is self-created, it is self-perceived, and it can be easily deceived as well in ways we can understand, predict, and ultimately influence during our lives.

If the world we perceive and believe in is a virtual reality, the product of a process with its own rules, then as that process undergoes predictable distortions from extreme stress, or in the stages of brain death, might we not then find ourselves in another universe entirely as real to us and just as believable as the one we now perceive? Is this what happens?

This is just one example of a new exploration which combines elements of religion, anthropology, developmental psychology, and developmental neurology. It opens the door on a new perspective, one which may change the way many people think about the way they think, know, feel, and even believe.

3

Painting By Numbers
Virtually Real

"One ought to know that on the one hand pleasure, joy, laughter, and games, and on the other grief, sorrow, discontent and dissatisfaction arise only from the brain. It is especially by it that we think, comprehend, distinguish the ugly from the beautiful, the bad from the good, the agreeable from the disagreeable..."

– Hippocrates

Reality, Appearance, and Perception

The world around us and our perception of it appear to take place at the same time. There is a real difference, however, between the world we are experiencing and the world that really is. To begin with, it all happened some time ago. It takes two hundred quadrillionths of a second for a rhodopsin molecule in the retina to swivel in place when it is impacted by a single photon of light. This momentary structural change initiates a series of interconnected events which eventually result in our sense of sight. This is where it all starts.

The photon is traveling, naturally, at the speed of light - but once it hits the rhodopsin, things start bogging down at the very first quadrillionth. Who knows, in other words, what's happened out there between the time the photon hits and the time we even start the process of seeing it? In some subatomic worlds, a lot can happen in two hundred quads. What this means is that, at any given moment, we are experiencing a world which is behind real time.

Returning to the eye, the signal now drops down through three layers of retinal cells like a Japanese pachinko game, hops aboard any of eight million optical nerve fibers, does a criss-cross through the lateral geniculate nucleus at the top of the brain, and arrives at the visual cortex as meaningless as an upside down Picasso. This means a further ride around the hippocampus to identify the image and assign levels of meaning and association.

Finally it clicks into consciousness. It's at least a tenth of a second later, and we finally see it, but the real world has already moved on. When any iPad can send a signal from Boston to Chicago and back in a millisecond, the time it takes us to see anything is a pretty slow train. After all, the entire universe expanded in a few seconds; humans are pretty slow at the picture show. Even worse, any image must mean something slightly different to each viewer. All our senses work the same way.

As a result of this necessary bucket brigade as the signals travel from one place to another, the world we experience is slightly behind "real time", it is a separately assembled and replayed version of a world that actually took place at some other time. In fact, any reality we perceive can only be such a representation, a replicated perception fashioned on the fly in our brain to which we react and relate as if it were real. It's our own "virtual reality." It can't be real, but we think it is.

This moment-to-moment process is seamless because we can't perceive the synthesis of perception. Moreover, the illusion is shared by all others. Since we share the same DNA and operate in the same manner, we seem to be sharing the world. In fact, we are all actually experiencing our personal show in our own cranial planetarium, and we never notice it. Even if we could, we could not change it and there would be no reason to do so. Still, as all our perceptions must be fashioned from a pattern of pulses traveling through neuronal networks, any reality we can perceive must occur subject to any number of limitations peculiar to our biological systems, just as a computer-generated virtual reality operates within the rule structures of its specific software

"Virtual reality," often shortened to "VR", usually refers to any "reality" created within a computer-generated environment. During the past few years, it has become the focus of expanding public interest in many areas. The trend has accelerated as the growing power of specialized processors support the newest advances in the rendering of realistic displays.

As usual, the US Defense Department Advanced Research Projects Agency (originally ARPA, now DARPA) was early on the scene. Beginning in the early nineteen seventies, work was initiated directed primarily at military and medical uses. By the nineties, a DARPA project was able to create complete incident databases by conducting interviews with every single participant of certain Persian Gulf battles. One result was interactive group-gropes involving dozens of trainees in video helmets blasting their way through unreal encounters of the virtual kind. It seemed to train them just as well as live fire and saved a lot of ammunition. It might also explain the surplus of "battle" video games on the market. Some gamers got a head start in virtual combat and liked it.

As the earliest virtual displays were designed for expensive surgical and aerospace applications, there was little popular use, and early attempts at popular virtual reality were a lot less than believable. Despite a flurry of films and speculation, early predictions fell short. By 2018, however, things had changed dramatically. With introduction of Oculus Rift and other mass market applications, including HTC's Avive, the Japanese KDDI, and Ali Baba-funded Magic Leap, the concept of immersing oneself in a virtual reality for any number of reasons had become the focus of billions of dollars of research, investment, and popular interest.

A typical observation was voiced by Boston's Andrew Miller, a vice president of engineering and self-professed digital geek. When asked in a Boston Globe interview what he considered his favorite app, he chose Google Cardboard, a clever cardboard virtual viewer that is both simple and affordable. "Virtual reality is something that has been stalled for years", he remarked," "Dropping the barrier to entry is the catalyst driving tons of innovation around how we can combine digital with the real world." By 2017, VR promoters had started teaming up with professional sports organizers to create choice virtual seating at any event.

Although current VR focuses on the visual, it would not be unreasonable, considering similar rapid advancements in other areas of computer science, to expect that within a few years we could enter a booth, adjust the goggles, put on a pair of transducer gloves, sink into a senso-lounger, and find ourselves in a jungle, on a beach, or on the surface of the moon. For five dollars a minute, we could live in a "virtual reality" battling tigers, exploring space, or romancing a movie star.

There are clearly a number of levels of reality at work in such a scenario. The first level is the only one the individual can, and should, perceive. Supposing, though, she were a coder familiar with typical limitations of the VR program. She might know her virtual Tom Cruise could respond well in a simulated conversation but probably could not hum. He could sing and harmonize, but only in major and minor keys. Any Bengali lass who wanted to sing a raga with virtual Tom would be mildly disappointed, but if demand presented itself, updates would follow in numerous scales. However, this is just the surface.

There is a series of ever more complex interactions at a number of levels working together supporting the Tom Cruise virtual reality, unseen by any participant but essential to the experience. The lowest level is the computer hardware. The next layer is the operating system, which organizes how the computer performs its tasks and how it interacts with programs and users; familiar examples are iOS for the iPhone and iPad or Windows for PCs. The program itself is written in a programming language, of which there are many, and each has unique rules and attributes that the programmer uses to determine "best fit" for the task at hand. A program will also require data in order to do anything useful. This can be anything from numbers to words, animations, video clips, music, and synthesized speech - even formats we have yet to envision. The program takes what it needs from various databases and merges it with new data collected from the user's interaction with the program. Finally, it must arrange all the parts into the same time frame and operate the various devices that bring the experience to the user. A VR adventurer slashing his way through a virtual jungle who happened to be a software engineer in the real world might at some point wonder how the program was constructed, but he would be limited to speculation unless he looked at the actual program code.

At the most fundamental level of any program is the computer's own virtual reality, the microcode itself. The basic computational environment of digital space is a binary universe, a world of zeroes and ones, the endless search for "signal" versus "noise," the "there" or "not there" of minute electrical pulses traveling at the speed of light through the murky chaotic static of an electromagnetic universe. In a digital processor, no matter how complex the command or the program instruction, information is ultimately expressed and manipulated as a series of ones and zeroes. Thankfully for everyone, the leap from "001001100111", which might mean "multiply value in memory location x by 2 and store in location y", to the wisps of fractal clouds drifting in front of the virtual moon on Tom's binary beach is simply too far removed for any meaningful correspondence. So the coder asks virtual Tom to kiss her and naturally he does. The surcharge appears on her credit card account. Finally, there are ultimate physical limitations which underlie everything else. A value is either zero or it is one; there is no one-half, or "maybe." There must be electricity and some form of memory. The silicon, metal, and plastic environment cannot be baked, burnt, broken, steamed, shocked, or boiled. If anything like that happens, the system simply won't work at all, and probably won't ever again. Goodbye Tom, goodbye beach.

This illustration describes a hierarchy of interdependent, ever more sophisticated rules and logical systems structuring and limiting all forms of virtual perception. This harmonious nesting of systems within systems creating the daily reality we perceive is very close to the Sanskrit term - *dharma* - a natural and continuing system. It is also why the impressive world around us is often referred to as *maya*, a word with three definitions seemingly at odds with one another: beauty, power, and illusion. We live in our dream, they say, entranced by the power of our beautiful illusion.

In fact there is no reality available to us except a perception which itself is a reproduction, created by a system over which we have little control and only the most basic understanding. There is no God in virtual Tom's world; everything is within a system which starts with a one or a zero. A similar series of hierarchies is at work within our brain at any moment, ordering, arranging, and directing a vast interconnected neurological environment, the unimaginably complex system required for the perception of human consciousness. By examining some of the systems underlying the experience of moment to moment perception, we may begin to determine, if nothing else, some of the basic foundations underlying rules that order all the days of our lives.

The biological demands of our neurological environment are simple and absolutely quantifiable. Our brain requires 3.3 ml of oxygen for every 100 grams of mass per minute and a blood glucose level of 80-120 mg per 100 ml. It must eliminate waste toxins, which requires a rich circulatory system. Every little part of it has precise requirements and limitations.

The brain can't survive ten minutes without oxygen at room temperature without sustaining serious damage. Any single blood vessel can clog or rupture for any reason, and irreparable functions can vanish in five minutes. We might never speak again. Within two hours a drop in glucose can be fatal, a threat too well known to diabetics. Anything at all that interrupts the blood flow will stop everything. The result is always coma followed by death. These are basic operating rules that cannot be altered; nobody has ever recovered from brain death. There are some other limitations, however, which are not so obvious.

Tricks in the Tapestry

The brilliant neurologist Charles Sherrington once referred to the brain as the "enchanted loom", as it appears to so effortlessly weave the seamless tapestry of our conscious experience. With modern neurological research progressing well beyond the basic biological mechanics of the brain, we are reaching the point where we can begin to describe both how the loom works, and how it doesn't work quite so well in some instances.

Many insights gained in this way have limited use. It is true we cannot perceive the ultraviolet spectrum or hear above twenty kilohertz, but the unheard and the unseen have little effect on everyday life. At a more basic level, however, nobody is suggesting that the brain is operating with anything but neurons. Whatever consciousness is, we perceive it with nerve cells and not muscle fibers. We also know that these nerve cells, if properly excited, propagate minute electrical pulses from one to another. Information is relayed by this means from cell to cell during normal brain activity. This is the brain's biological electrochemistry in action; it is not debated as a theological point or a philosophical conjecture.

Since a pulse from an activated neuron cell is the equivalent of "1," while the latency period without a pulse is the equivalent of a "0," the most basic underlying operating imperative of our perception would be the ability to sense the difference between the two. Our consciousness seems to arise from a sophisticated form of chaotic pattern recognition, and a pattern can be defined only by the use of contrast. To that extent, brain cells and other digital systems share a common reliance on comparative functions to get the computational job done, the same signal-to-noise ratio, pulse or no-pulse, dot or dash, "there" or "not there." A neuron that couldn't tell the difference would be as useless as a binary circuit that didn't know zero from one.

At the surface level of our awareness, this most basic functional operative is completely invisible; it does not affect the colors of the day or the thoughts of our mind. Only if we try to think about something that a pulse-based consciousness cannot think about do we get into any trouble. Usually, when we try to do this, it can be difficult or even disturbing, almost as if there were something wrong with our mental focus button. Try, for example, to picture "forever". Or "perfect". An incomparable abstract simply won't fit in a comparative cognitive environment. We know what the word means, but abstracts are slippery. We can't make a mental picture or find words as we can for images we acquire from experience.

"Forever" is just one example. Human consciousness is perceived through neural communications in which everything depends on the presence or absence of voltage potentials. Since our method of cognition is comparison, we can't communicate about any non-comparable states at all. We can't really describe "perfect" any better than we can paint "never"; it's a problem with information being passed around in a pulse-based form. This is not to say that we cannot have experiences or feelings we could call a perfect moment; it's just that we cannot use standard cognitive reflective thought to describe or articulate them very well.

In fact, events which evoke non-comparative expressions are often touched with personal or even religious overtones. At such moments, few would suggest that normal reflective thought was predominating. In many ways our hormonally-driven emotional feelings resemble wave phenomena, more closely related to analog events, while communication and cognition tend toward the digital. The whole of our neural electrochemistry embraces both levels, but we "think" in ways we can discuss with other people and "experience" those non-comparative states that we cannot ever really communicate in a rational or cognitive manner.

Perhaps, then, we cannot know the nature of God simply because neuron-based brains can't handle the infinite, per se, at all. From a cost accountant's point of view, this makes very good evolutionary sense as we don't encounter that many incomparable beings in our lifetimes. It could also be one good reason why it has been so difficult to communicate with the divine, or at least why it might be hard for humans. It seems the incomparable can happen but it doesn't compute, just as we can "know" and "experience" things we can't think about or describe in words.

This is but one example of the perspective required if we are going to harmonize the truth of religion with the truth of science. There is no scientific problem with saying "The perfection of God is hidden from the understanding of man" because, neurologically speaking, the human brain can't really mentally image a "perfect" state, whatever it may be. It is an inherent design limitation of a neurological method of perceiving consciousness, and we wouldn't be humans without it. The faithful would say God made us this way, while others might blame molecular genetics, but in either case we can't think of an alternative. Not with the neural equipment we've got, anyway. Who knows? It doesn't really much matter. We are interacting with a chaotic world every day, and it's a blessing that our consciousness does as well as it does, even if we can't see ultraviolet or describe the transcendental.

This sort of viewpoint can provide a space both for the experience of the divine from a personal point of view - for those who have had such experiences and cannot deny them - and for a scientific, rational explanation as well. God alone could divine the basis and method of divine perception. The basis of human perception remains locked into the systems within systems of brain cells that pulse or don't pulse, require sugar and oxygen to survive, and cannot be damaged or starved or they will die and, of course, take us with them.

Time and Space Taffy

At the most basic level, we know that human conscious perception arises through the interaction of neurological systems functioning in a delicately balanced biological environment. This perspective goes much further than simply detailing the limits of the human brain. Extrapolated into either religion or science, it can be equally insightful. On one hand it may dim the attraction of seeking perfection if perfection is inherently personal and we can't describe it using the human mental system. Clearly, if we can't adequately identify it, we certainly couldn't tell anyone else what to look for. It seems we may have to "know" it when and if we find it, somewhere in the realm of personal experience, both unrecognizable to anyone but ourselves and dependent on the circumstances of the moment.

On the other hand, science fares no better. Norman Ramsey, who won the Nobel Prize in physics for developing the atomic clock, pointed out at a lecture in the Harvard Science Center that time is measured by periodicity. From sunrise and seasons to the picosecond beats of vibrating cesium atoms, it is recognition of the regular repetition of something that sets the foundation for the perception and measurement of time. Unfortunately, perceiving the repetition of anything at all requires a conscious chronological memory, apparently reserved exclusively for humans past the age of two on this planet. Moreover, it seems our perception of time stays linked with that of those around us only if that periodicity remains constant.

The problem with a biological basis for time perception is that neurons aren't as tough as silicon or cesium. In humans, things happen. The body's response to sudden stimulating events is immediate release of powerful hormones, one effect of which is to dramatically speed up neural firing rates in certain higher brain areas. It makes excellent sense to increase the speed of data processing when something really exciting or dangerous may be happening. Neural firing can speed up by three hundred percent.

The problem this causes with perception is that gulping down information faster makes the world around us seem to slow down. This time-expanding effect, described in greater detail in later chapters, is similar to speeding up a movie camera in order to create slow motion. If parts of the brain get out of synchrony, all sorts of weird things start happening.

Since this suggests that any individual's perception of time must be flexible, it could give even Einstein a headache. If time and space are interdependent, but time is a variable relative to the observer, is space, then, also a variable? Can we experience infinite space like a moment of timelessness? If we were to oxygen-deprive the brain, are we in another time and space or are we in brain failure? How can we ever really know anything for certain unless our knower is standardized, yet we know full well that all human brains differ slightly from each other?

Worse, biochemically speaking, at the molecular level, every thought must change our brain chemistry a little. Heisenberg's classic uncertainty principle states that since we can't observe anything without pushing it a little with whatever we find it with, even a photon, we never really locate anything exactly because we just moved it by finding it. If we can't think about anything without modifying the biochemistry a bit each time we think, what does that have to say about the nature of any search for ultimate answers? Wouldn't the questions change as the questioner changed in the process of working out the answers? Or is our mental journey the answer itself?

Questions like this are bound to arise as we begin to explore some of the operational aspects of our biological lens of perception. The answers do not deny religion; but just as the incomparable is unthinkable, the sense of time itself is probably a fairly recently evolved capability. If we could neither recall nor project very well, we might not ever think about what happens after death until it was far too late. In fact it seems highly likely that most, if not all, of those hard questions requiring religious answers may have been nearly impossible to conceptualize as little as 100,000 years ago. Recent discoveries indicate that the ability to sequence time, generate abstract thought, and speak coherently all require brain structures evolved within a relatively recent evolutionary time frame. The earliest humans with larynxes like ours, for instance, didn't appear until about 200,000 BCE. Did we chip hand axes for half a million years without any conversation beyond a grunt? Nobody's said a word.

More to the point, for Adam to understand God's commands, or even speak with Eve, he needed a well developed speech cortex. Clearly, any Eden had to appear at least past that point in our evolution. In fact, without many recently evolved neurological capabilities, we would not have had the consciousness to know either natural law or divine intent; and even if we did, we certainly could not have written about it, read about it, or even spoken about it to anyone else. As far as our world religions themselves are concerned, not one is more than 5,000 years old; actually a rather recent phenomena, just as we ourselves are.

Everything in 1,400 cc's

When it comes to higher forms of life, consciousness is usually characterized by the relative sophistication of perception, the ability to extract information from the environment, and cognition, the manner in which that information triggers a useful reaction. Limiting either limits our experience. Without perception we would have nothing to react to, and unless it were for some purpose, we would have no reason to react. When the manipulation of perception and memory to some useful end includes the use of abstract thought and projection, we call it reasoning. It is the higher levels of reasoning and perception that we call intelligence.

When it comes to judging the consciousness of another creature it is useful to remember these variables. The honeybee, for instance, perceives and is conscious of ultraviolet light. It can see colors we cannot imagine or know in any way. On the other hand, a bee's tiny "brain" is too limited for any cognitive processing. It cannot adapt or decide anything consciously. It cannot reason at all. Moreover, its minimal insect consciousness must act through a nervous system of great simplicity and efficiency, and the bee is further limited by this simplicity. Insects are practically hard-wired, entirely pre-programmed. If a bee heading in a bee line meets a breeze, increased air pressure on one side of its body automatically energizes a muscle linkage that angles the beating wings, like helicopter rotors, to compensate for the sideways drift. The bee doesn't know it happened. Moreover, its multi-lens eye can't focus, so it nearly runs into flowers.

Returning with nectar or pollen, the bee dances directions to the flowers, turning in patterns on the hive wall as the other bees brush up to get the latest travel reports. It would be nice to imagine that bees are scrupulously honest insects, since not once has a bad bee knowingly passed on false information. In fact, they can't. It's the playback of the flight recorder operating the bee, turning the insect into a dancing marionette mindlessly miming something it can never understand. Aside from lacking alternatives, insect brains have little internal redundancy because insects wear out before they need replacement parts. Four generations of honey bees live and die during the time it takes one human brain to mature - four years during which our unique human consciousness acquires complexities and capabilities we will never fully understand.

To know the soul, it must be perceived by a human consciousness for examination and reflection. For love to guide us, we must be conscious and aware of love. If we are unaware of ourselves or others, we are called thoughtless. When we act without the awareness available to us, we are not being mindful. It is our consciousness alone that makes the universe known to us; and our consciousness is made known to us only through the moment to moment functioning of our living brain.

Within the space of roughly fourteen hundred cubic centimeters moves the exquisite organic instrument which determines our entire awareness of anything else at all. Damage it or stress it and we are no longer aware of anything in the same manner. Our universe will change around us. We may have a change of heart, we can change our minds, and the brain will accommodate our shifting realities without a missing a beat. But if we tamper with any basic function of the brain, we can distort or destroy the perception, realization, and projection of our entire consciousness for some time, possibly for all time. Our world, as we know it, is in our hands. More precisely, it is in our heads.

It is here where the paradox of a physical brain and a non-physical mind, spirit, or soul may at least be partially resolved. Whether our consciousness, with which we perceive everything else, is created by the brain, or simply perceived by the brain, it still can only be as we perceive it. We perceive all of it through the structure and the function of the most complex and intricately arranged form of living matter we could know or imagine. It must function at a level beyond description since it must be complex enough to let us perceive anything we can describe or feel, as well as regulate and operate everything else at the same time. It is as close to the infinite as we can get close to, and it is not out there. It is in us, a part of us that makes us who we are and what we are, and it is alive and well or you wouldn't be reading this.

It makes for us the only world we know, a physical activity that lets us perceive our days, our nights, our dreams, our faith, our beliefs, and any other thing we can perceive at all. Whenever we use our mind to search for meaning, we will find it wherever we look the hardest, and we will always find it to be made of whatever we believe in the most. If we sense that the universe is in a constant state of creation and change, it may be because we perceive it with a living mind, born of a living brain, itself in a constant state of creation and change, a daily ongoing enterprise of billions of minute living cells.

Cells to Think With: The Binary Brain

Human neurons are complex, efficient, and interconnected at a level of sophistication not found in most other living creatures. Even observing one neuron going about its solitary business is to witness extraordinarily complex activity. Aside from the life support functions of metabolizing glucose and oxygen and housekeeping all sorts of chemicals and hormones, each cell is always communicating with hundreds, even thousands of others.

Each neuron has a unique voltage threshold. Constantly receiving pulses from thousands of other neurons, several times a second it sends its own pulse to its network of interconnected neighbors, a constant chattering of excitatory or inhibitory pulses. Any time a neuron's internal voltage (excitatory pulses minus inhibitory pulses) rises high enough, the neuron itself will "fire," sending its own electrical signal downstream to join the chorus of thousands. If it gets a lot of sudden activity, it shortens the time period and sends the signals along in faster bursts to accommodate any temporary data overloads. Neurons do more than simply relay signals, they mitigate hormonal responses as well as perform other functions; it's always a team effort. It is useful to note the resemblance between a neuron and a tiny digital processor. It has its internal instructions to fire when it passes a certain voltage level, and only then. Its life is endless cycles of averaging, pulsing or not pulsing; adding, subtracting, and calculating like a patient little accountant living on air and Twinkies. Several times a second it says "something" or "nothing," and resets for a new cycle.

This unfortunately leaves us with the inescapable conclusion that at some level every thought, feeling, and perception that has been either stored or realized is an immense, unfathomably complex pattern of voltage potentials. Not only is this hardly poetic, it conjures up images of analog to digital interfaces separating the harmonious flow of life into binary bits like some vegetable-chopping appliance. Once again we are confronting the popular image of the brain as the ultimate digital computer. Still, it makes perfect sense, as well as complying with Occam, that the brain would have by now evolved to the most efficient way of doing its work. There are plenty of good reasons for using a binary system.

The underlying basis for binary in computers is that as long as one has sufficient speed, it is immaterial whether the computer sees the number 357 as a sequence of the three Arabic numerals or as a long string of ones and zeroes. If the mechanism, or the organism, only has to distinguish between two possible states, it's that much easier to identify a signal against the background "noise" always present in a calculating environment, be it silicon or cellular. Everything can be simpler and more efficient. Since computers are so exceptionally quick, they don't mind working in digital and they have been doing so from the very beginning.

We had a typesetting computer back in the dawn of such things, when even personal computers were still ten years in the future. It was dumber than most pocket calculators; all it really did was convert typed code into punched tape to be read by a photo-typesetter. It had a program with about 800 steps, the functional IQ of a mollusk, but it was impressive for its time. The poor idiot had to go through its entire program every time it wanted to do anything. Everything it knew was in 800 consecutive steps, a one way smorgasbord-on-a-track that allowed no deviation.

Everything it knew was in those 800 steps, but it didn't have the sense to go looking for anything in particular. When we typed "A", in the first cycle it noted that an "a" had been sent from the keyboard. It registered "a" in a little cache memory and raced through all 800 steps again to find out what this "a" was going to turn into. Then it picked up the tape punch code and jogged through the whole shebang a third time to tell the punch what to do. IBM made the keyboard, Photon made the computer, and Litton Industries made the tape puncher. If anything went wrong, which was not uncommon in those neolithic times, repair wallahs from several billion dollars worth of corporations would show up and blame each other as old stupido ran around in 800-step circles.

Fortunately, solid state silicon switches are very fast. So fast, in fact, that our poky little computer would run through all 800 steps about a thousand times a second. It was a manic whirring electronic conveyor belt, hungry for digital bits. With a small "zap", the "A" was on the tape before we could get to the next letter. Any device that gobbles pulses by the microsecond doesn't mind if 357 comes in threes, or in three hundred ones and zeroes. It has all the time in the world; a relaxed sort of digital reality.

Human brain speed is relatively slow, rarely exceeding eighty miles per hour, but it more than makes up for this speed limit with massive redundancy and complexity. Until recently, most processors used the model developed by John von Neumann in which computational steps occurred sequentially. From relics like old stupido to the early multi-million-dollar Crays that zipped along at billions of operations per second, programs progressed step by step in an orderly, if frantic, pace. Advances in technology made possible new generations of computers built on another plan. These "massively parallel" designs employ specialized processors to separate computational tasks and feed them to a swarm of hungry little sub-processors all at the same time. The new designs are multi-tracked and much faster. The human brain does it one step better. Composed of over 80 billion neurons in a three dimensional hyper-connected matrix, it is both massively parallel and three dimensional as well. "Unimaginably intricate" is both an accurate description and an understatement.

If we are to send information around in a complex structure, and the brain is infinitely more complex than any computer, it would be helpful to use the simplest codes possible. The binary system makes for far less confusion; pulses are simply "there" or "not there." Ultimately, if these multiple, multiplexed, interwoven codes are complex enough, we can express nearly anything. As a result, all of our senses, both internal and external, send their information into the brain coded into a string of pulses. From the taste buds on the tongue to the tone receptor hairs in the inner ear, almost everything comes to mind initially as a pattern of ones and zeroes. It is the major business of the brain to integrate this information sequentially with any and all pertinent information available in memory and react to it, incidentally creating this grand virtual reality we call the experience of life.

It might seem impossible at first that a consciousness such as ours could be adequately perceived through something as simple as a molecular Morse code, but it is not as difficult as it may appear. As an illustration, we can examine how our brain lets us "see" a sunset.

Colors from Zero to One

First, we have to learn to digitize visual information. Say, for example, we're going to transmit a picture of a tree, a ten-foot by ten-foot mosaic for a garden patio. We have only black stones and white stones and they're about the size of quarters, about an inch in diameter. With a ten-foot by ten-foot area, we could get a good representation of a tree shape with a patio-sized grid of 12 lines to a foot, a total of 14,400 stone-sized black or white stones making up the entire scene. It's going to take some time.

As it happens, an acquaintance hears of our artwork but his only connection to us is a one wire telegraph. One day he taps, in code, "Have stones, have grid, send pattern." A pulse will mean a black stone and an equal period of time with no pulse will mean a white stone. As long as the system is understood, we can send 120 sets of 120 pulse/no-pulse strings. Our friend can then reproduce our mosaic perfectly, stone by stone. If we want to make the picture more detailed and subtle, we would only have to increase the number of stones in the grid. At some point, greater detail would require a patio the size of a football field, so we could reduce the size of the stones instead. Retaining the ten-foot by ten-foot format and using pebbles the size of tack heads, we are expanding the grid to 1,200 lines to a side. With that sort of delicacy, we should be able to produce more life-like pictures with a number of gray shades when viewed from any distance. Still, the entire composition could be sent on one wire with zeroes and ones.

By the same token we notice lasers, inkjets, and printing plates doing it the same way, all using lines of dots or no-dots. Daily newspapers employ a 65-line-to-the-inch grid, enough to produce gray shades at a few inches. At any time we choose, however, the picture can be unraveled into its strings of ones and zeroes, sent through a smart phone and be reproduced perfectly by a printer anywhere. Full-scale color printing isn't much more complex. Any visible color can be created with the three basic printing colors cyan (blue), magenta (red), and yellow, plus black - the same color inks found on ink jet printers.

For commercial printing, a color original is scanned with a laser beam. This allows the image to be broken up into lines of pixels registered by photoreceptors, each sampled for the three basic colors. This information is fed to another laser that then scans four separate negatives, line by line, creating four grids of tiny dots. The photo is now "separated" into its component colors, with one dot-grid negative for each color. From these, four printing plates are prepared. This is why it is called "four-color process" even though many more colors will be represented in the end product.

The paper is then printed sequentially with the yellow, blue, red, and black grids. If the original color was green, there are tiny dots on both the blue and the yellow printing plates at the same place, blurring to green as we look at it. Small dots from the black plate create a darker green where a shadow falls over a leaf in the original photo. For shorter runs, high-tech commercial ink-jet printers fire tiny bursts of color at extraordinary speeds as ribbons of pages shoot under rows of print heads.

This is also the way we saw the delicately colored pink and orange bands on the planet Jupiter, the icy blue of Neptune, and the red hills of Mars. Instead of printing the colors, the photoreceptors on the spacecraft transmitted their binary strings directly to Earth, where they were used to form colors on video screens. It was all ones and zeroes again, with slightly more complex breaker codes that switched scanning lenses and software. These days, the impracticality of shipping paper long distances has resulted in the printing of national magazines at printing plants located in different geographic areas. Using the same binary codes, brilliant images are flashed from coast to coast in seconds and still look exactly the same wherever they are reproduced, printed, and sold.

When it comes to scanning photoreceptors, the human eye is without question unique in the animal kingdom. We share the gift of color sight with very few other creatures; every dog has his days, but they're all in murky greens, browns, and grays. Aside from the ability to see over 30,000 shades of color, one of the most amazing things about the human eye is its ability to handle grada-

Painting By Numbers 59

tions in color and brightness from bright sunlight to shadow without altering color values; it never has to change film or rely on filters. We are also blessed with true stereoscopic three-dimensional vision, but most importantly, we do much more with it than any other beast or bird. The wild turkey has a much more accurate eye than we do, but they're real turkeys when it comes to the thinking part. In the human brain, areas of the visual cortex which interpret data from the right eye are physically interwoven with the left eye's areas in natural patterns that resemble a fingerprint or a zebra's stripes. Between our optics and the interpretive ability of the various layers of the visual cortex, the human 3-D color-correcting sense of sight is the number one picture show on earth. Still, it is all in binary code.

The cone cells in the retina register blue, blue-green, and red light. The rod cells, used mainly for low-light vision, register only black and white. By a complex process known as color subtraction, those three colors do the same job as the four basic printing colors, plus black and white. The retina's layers contain several levels of specialized cells to register edges and movement as well. With the delicate muscles and the lens of the human eye to direct and focus images on the retina, we have a natural grid with which we can register any visual image between the infrared and ultraviolet ranges.

How subtle should we get? As it happens, each eye has about 120 million retinal cells. With top of the line cameras offering 20 megapixels, it's difficult to imagine a camera with roughly 120 megapixels, but that's what we have at our disposal; it's a pity to waste it on black and white type. Looking very closely at color printing, most of us can make out the color dots at 120 lines to the inch. *National Geographic* likes to be special, and prints its color at 180 lines to the inch on its own presses. When we pass 600 lines to the inch, the eye cannot tell printing from photography. At twelve million lines to the inch, we can't tell perception from reality; will never detect the retinal mosaic. The digitized patterns that appear in the mental theatre of the mind's eye are seamless and totally believable. We think we see directly with our eyes.

Perhaps even more startling was the observation that nerve fibers going from the brain to the eye are just as numerous. This suggests that, employing the same practicality of all video screens, our own visual screen may not be constantly re-created in its entirety. Why not save energy? Simply re-play anything that's not moving from memory, and only re-image whatever's changing.

That would save a lot of brain cells, but it also means whatever we're looking at that isn't moving might just be an echo. The imaging system works fine even when there's nothing to see; our dreams never fail to fool us. Even dogs paw in their sleep as visuals recorded at other times play back through their closed eyes. Is our world all there, or is much of it a memory of the way we remember it, and it just hasn't moved recently?

Back to the eye, massively parallel retinal grids now pour their information down through the three layers of interconnected cells to further define shapes, shades, and shadows. This results in some data compression, but the resulting chorus is now funneled into one of the eight million long optic nerves, each chattering away in strings of pulses up to a hundred times a second. This is all happening long before we even know we saw it.

At twenty-four frames per second we watch films, and at thirty frames a second we watch television. Our visual images, also, are reproduced frame by frame at the very back of the brain. The neural activity then seems to wash forward, picking up meaning and context from memory in brain structures downstream. Suffering a stroke here, the victim might see perfectly well but might not be able to make sense of it. When everything is working, however, we are presented with a nearly instant view of the world we know, reinforced by our memories and deep with meaning. It has taken a fraction of a second, many millions of cells, and hundreds of interlocked neurological systems. Human sight is so much more than a cellular camera.

A Colorful Line of Thought: Synthesis and Sunsets

So now let's use our inner vision to imagine we are seated on a bluff overlooking a rocky California beach a little north of Santa Cruz, gazing out over the Pacific. It's a warm Sunday afternoon, fading to the last part of the day. It's been hotter than we expected; we notice a little sunburn on the neck, now, with the sun low on the horizon. The warmth lingers, but the breeze is picking up. With the day cooling off, it's time to just relax and sit on the grass and watch the sun go down.

There were showers in the afternoon, and a last flock of dark clouds are scudding slowly off toward the west, blocking the sun while letting its dying rays pierce through here and there as it sinks towards the sea. Then, for a moment, the lower edge of the sun begins to emerge slowly from the bottom of the lowest cloud, glowing at the edge of the sea, suddenly brightening the horizon and bathing the bluffs and waving sea grasses in that unique horizontal yellow light that blazes out when the heavens are dark and the sun is coming from a crack at the edge of the world.

For a moment the sun rests there, suspended, glowing in deep oranges, and then slowly sinks into the sea. The waves hiss over the sand as the twilight descends, the pink cotton candy clouds rolling to magenta and fading in gentle deep purples. Shadows begin to wrap the rocks in deepening darkness while in the east, the silver slice of a crescent moon, shining against the cobalt blue sky, begins its climb towards an evening star. The breeze is getting a little chilly now, and it's time to for us to getup and head back toward the house, the windows alight from inside, glowing against the last twilight of a soft evening as a quiet night slowly cloaks the shore.

The brain remains in silence and in darkness. Sixteen million fibers are pouring cataracts of information over an infinite grid as our mind fills with the sunset, and we are surrounded by it in all ways. We can never be aware of those billions and billions of ones and zeroes, we can never hope to see them although they outnumber the stars in the sky. It all happens so fast and so neatly that we see a real sunset, and only that sunset, in depth and color possible only for our human eyes to perceive, and a beauty only our human mind could know.

The sun has set; the grass crunches underfoot as we walk up the path to the cottage on the bluff. The screen door slams; Annie's dress waves on the clothesline behind the kitchen. Note on the table - "Gone for pizza, back soon." The cat purrs by; it's time for night stalkers to awake. Cats have reflective retinas with nearly nightscope sensitivity after dark. Perhaps the cat savors another few minutes of sunset? A feline is quiet. Who knows what she thinks? She thinks in her binary code, her tail arched over her back, as the last light fades from the sky.

Part Two: The Past

4

In the Beginning
From Heaven to Earth

The ovum is pierced. Genetic traditions from all our family ancestors spill into each other. Dancing chains of ribosomes, jeweled necklaces of life, embrace and entwine. Hesitant groupings of characteristics from both sides extend atomic greetings, holding tight with the clasp of phosphate bonds; twining, twirling, fusing. Now we are. We know nothing - and everything - because we are all that we could know. We are the one and only because we are the only one; a one-cell dream of a future self. In that endless moment, the pulling and combining and joining are making us, and taking us, from never before until forever after. Here it all begins. We are weaving into something that was never before, nor will ever be again, but is here now, and very new. Still in a time of timelessness, the fertile cell divides, and divides, and divides again. Patches of genes awake with specific organizing powers. The entire composition is re-recorded in every cell, complete with all assembly instructions. Here will be the feet, here will be the eyes, and here will be the brain. In the eternal darkness, our home is forming, and we are forming, and nothing is left to chance. There is a place and time for every part of us, and we grow.

And where is the mind? Will it reside in the toes? But those who lost toes to the frost became wiser for the experience. Was it nestled in our budding heart? Many hearts have been traded by transplant and there has been no sharing of the spirit. If our world is to be perceived through the brain during our life on earth, it will have to exist in a very limited form for a while. The eyes are not finished yet, and there is nothing to see anyway. There is no place to store a memory; those abilities will come much later. But we're here. We are here from the very moment that the joyous dance of life begins, unalterably and completely ourselves and only ourselves from the moment of our creation, long before we have enough of a mind to think about life. We are woven into every strand, and from this point on, we simply locate our cells, find a place to settle down, learn to specialize, and multiply.

Now comes the long and endless sleep of quiet building when currents and connections less thoughtful than thought and many times more profound are forming that exquisite part of us that will allow us to perceive our life and introduce us to the world. We have nearly nine months to go. Nine months to create, bit by bit, the biological basis for a consciousness that will, one day, perceive our spirit, our mind, and our soul. Like all truly beautiful expressions of nature, it takes time to form and grow, a time to awaken us to the world we know, and finally a time to reach its end, and leave us timeless once again.

In our embryonic brain, our first perception, our original mind, is oneness, and only oneness. There is no time that can compare with this, because with only one there is no comparison. The time of oneness is always forever; and then the cell divides and we start the time of two. And then comes the time of three, and the time of four. As each new living neuron comes into being, our growing brain becomes, by that degree, more discriminating. By the time we are three months along, our brain is adding neurons at the rate of 250,000 per minute; by the time of our birth it contains nearly a hundred billion of the most complex cells in the body. It is more elegantly specialized and balanced than anything in the universe known to mankind, as it must make that universe known to us and balance our life within it.

We remember none of it. We can never remember when we were all female, for instance. It is not until the third month that the male fetus produces the hormones that alter his body and brain to make him male. Males can be feminized, and females masculinized, by abnormalities in a mother's hormone ratios at this crucial time in brain development. Even severe stress during pregnancy has been linked to gender-related anomalies. A pregnant woman requires emotional as well as physical well-being as her child's mind forms, moment by moment in her womb. Likewise, none of us remember when being and knowing were the same even though it seemed to last forever, since we couldn't tell past from present. In our own endlessness we were moving steadily toward a meeting with a world which, having never been experienced, we could never have imagined, the world that we call time and space, where we will spend some time and take our place.

There is a continuing controversy as to when we are a person. Some believe it to be at the moment of conception; others wait until the fetus can survive outside the mother. All seem to agree that the child, once born, is a very small, very young human being. But it is not a finished human being. The passage down the birth canal is not the final stage for any part of us. It is only a physical interruption in our maturing process that transfers us out of our mother when we can survive in the world outside. Survival was one of those things we hadn't thought about the week before it happened. None of us expect to be born; we all expect to remain in eternity forever. In fact, at that point we don't really expect anything at all.

This is our beginning, and this is also our ending; this is the eternal place we must leave in order some day to return. Nearly every cultural myth of the creation of mankind is just a broadly interpreted description of the experience of birth from the point of view of an infant being born. We had always been in no time, no space, all time, all space; our only name was "I am" and we always were. In fact, we were about to take human form, transformed to an infant in a mother's arms. It was a blessed event, but it was also a bewildering one.

The Creation Story

It is dark, with a dull redness during the day. The fetus is moving its limbs, drifting and turning in a personal universe, growing more and more aware, yet all-knowing and all-being. By the sixth month, we can already smell and taste and lately there have been sounds as well, the murmurings of the gods are getting clear and clearer. By the seventh month we can distinguish voices. Through it all, the heartbeat of a world still shared with our mother is imprinting us with the pulse of life. From the lives before our own, beyond the beginnings of time, this is the rhythm we will always seek, will always be calmed by, and will even sway to if we feel stressed, rolling back to our very beginning and the wordless prayer we all know by heart.

For a time now, we have become aware of changes. In our eternal darkness, a new spirit moves over the waters. Suddenly there are great movements, voices become clearer. The creator is about to start up the world for us and the powerful contractions begin. The obstetrician said "Turn up the light," and there was light, and she saw that you were good. It didn't take seven days either, but it is a pretty chaotic experience so it could have easily seemed that way as we tried to makes some sense of it. And what a demotion! We had been the entire universe, the be-all, he-all, she-all, and end-all. Now we were just as helpless as a newborn.

They turned on the lights, the crowd cheered, and it was a whole new ball game. We called foul. We cried. We yelled. We were really put out; figuratively, metaphysically, and literally. Our mind wasn't started at birth, but we must have been startled. Drugged with natural endorphins, we were shoved down that dark tunnel into the blinding light. It had been forever in stage one, and then we were gasping and blinking and kicking our way into this new stage, and suddenly we were center stage. All newborns must dream a lot about the old days, wondering about it all. They spend nearly half their time in REM sleep, the dream state, even with their eyes open. They just can't believe it. Forever and ever - and now this utter confusion? What happened?

We keep waking to a new reality, and we cry a lot about it. You can't remember; no one can. We talked in baby talk, and we thought in baby thought; we can't recall anything very specifically because our baby brain was still so non-specific. Creatures that operate entirely on instinct, practically up to the reptiles, do arrive ready-made. Just hatch them, and they're up and running. Aside from their size, they are fully wise and capable, as wise as they are capable of being, from their first days on earth. Here they come, and off they go.

More complex creatures take more time to mature, and we mature gradually as our parts mature. We come onto this earth both unfinished and unorganized; we can't even eat solid food for many months. There was no part of us that was, at birth, fully detailed or final. Every part was infantile; baby toes, baby nose, baby fingers, and baby brain. Everything was already working, or we could not have been born alive, but there's a long time between appearance and maturity. Every part of us had years to go. Our brain was far from being organized and structured as it is now: articulate, differentiated and working with years of experiential memory. It was baby's brain, and it was about as capable of reflective thought as baby legs are for running. It still had to develop further, and all the time it was perceiving and understanding the best it could with what little it had. Given our baby legs, we would stumble and fall; we were not ready for gravity then. With our baby brain, our consciousness was equally incapable of the sure and distinctive method of thought which characterizes the adult mind. We were much closer to forever then than we are now.

There are several interesting aspects to the stages in which the brain matures. We are born with nearly all our neurons, and these cells rarely reproduce. For reasons which will become clear, it would be impractical to deal with the constant appearance of too many blank, immature, or disconnected cells in the midst of things. Instead there is enormous redundancy. Even after a stroke, there are often enough spare neurons around to eventually re-connect and re-learn. With billions of cells, we can afford to lose a couple of thousand a day all our lives, which we do, and yet not run out during our life. Between birth and the age of about two-and-a-half, each of us a little differently, we are navigating with a consciousness that is constantly on the run as our brain hooks itself up and prunes itself down to the right size for a lifetime career in data processing.

We know that each neuron communicates with countless others, sending electrical pulses down an exit nerve fiber, the axon. The axon in turn splits off into numerous hair-like dendrites, tiny sub-fibers. An axon which has grown all its dendrites is said to be fully "arborated," from *arbor*, the Latin word for tree. It looks exactly like a tree without leaves, dividing and subdividing from major branches to the tiniest of twigs. In this way, a single nerve cell in the brain may be in contact with up to as many as 10,000 others. With nearly all these cells in place at birth, much of the next three years is spent in the gradual arborization of the axons and dendrites. Our chips were in place, but they weren't wired up. We have to make our connections before we can make our communications.

By the time we are one year old, consciousness is undergoing a very significant change. At first, it took lot of energy to push impulses down those innumerable pathways. Within a few months, however, specialized Schwann cells start to wrap each axon in a fatty layer of insulation called myelin. This allows electrical impulses to race along nearly ten times faster while using far less energy. An often overlooked benefit of this upgrade is that our large human brain takes a quarter of the metabolic energy we produce, and yet runs on only a few watts of power. With the myelin retrofit, the brain is soon up to speed, rapid, efficient, and getting ready for the complex micro-movements which will let baby take his first steps.

It is during this time of myelination, as the process is called, that malnutrition can cause mental retardation in an infant, a primary basis for promoting early childhood nutrition. The infant brain is still very vulnerable. From a virtual reality perspective, however, the world we'd just gotten to know must have really done a flip as we supercharged our basic operating system on the fly. We completely alter the pace and the perspective of perception, and yet we seem to do it imperceptibly. That is to say, no infant has ever noted the transition of reality from what we might call our "universal archetype mythology" state to the "ancient real memory" state. The technical term is "infantile amnesia," but it has much less to do with forgetfulness than with our rapidly changing perspective in a timeless environment.

The proliferation of dendrites during this time is profuse, creating yet another effect on perception. No matter how a memory is recalled, it must be stored in some place that won't change significantly so it can be retrieved accurately. Complex memories would require either huge storage spaces or enough little storage areas to hold as much detail as necessary. Luckily, the complex arborization of human neurons makes this task possible. The brain never runs out of complexity. At maturity, with tens of billions of cells hooked up to thousands of others and each capable of a nearly infinite number of electrochemical energy levels, there is more than enough variety for a memory large enough to let us learn new tricks at almost any age. But there are tricks that the growing mind plays on us while we are still infants.

For years after we are born, each neuron is growing more complex. Year after year, dendrites split, divide, and grow even as we are reacting to the many aspects of our environment, finally reaching their ordained places and settling in for many long years of electrochemical exercises. The brain actually reaches its greatest internal complexity at about the age of three and a half months -

but as we continue to mature, a certain percentage of less-active dendrites die off.

This final "pruning", as it is called, continues into early childhood. It leaves us with our basic neural networks, our unique lens of consciousness through which we will perceive our hopes, our thoughts, and our world for the rest of our lives. There has been recent speculation, elaborated in a 2014 study by neuroscientists at Columbia University Medical Center, that children and adolescents with autism suffer from a surplus of synapses in the brain due to a slowdown in the pruning process at this stage in their development. Because synapses are the points at which neurons connect and communicate with each other, too many connections may have profound effects on how the brain functions. Fortunately, most of us mature normally. As our activities strengthen and reinforce our neural growth and interconnection, we begin to express and reinforce those basic mental idiosyncrasies which will mature over time into both our personality and our entire underlying image of the world.

By the age of two and a half, the major biochemical and structural upgrades are complete, although final maturation progresses slowly through adolescence. The rapid growth phase is now over, and the network is stabilized. Memories are no longer distorted or transformed by brain growth; they can be chronologically associated, retrieved, and recollected. With the pattern-sequencing capabilities of the prefrontal cortex finally coming online, the sense of time is becoming distinct. Children now consciously differentiate; they know they are little boys and little girls. The fresh mind is ready for training as we begin to learn from our own recollection of day-to-day living. All of us, in all lands, in all families, slowly become self-conscious, and socialization begins. We are not center stage any longer, but among others. We are coming into contact and context with the world around us and every day more out of touch with that eternal world that was ours ages ago.

Only before we could perceive ourselves in context could we be ourselves in essence. We had been like that for such a long time, with birth itself just a major incident. For years afterward, the world turns in sympathy with the churning activity of our baby brain as we weave our way to self-conscious thought. We enter this world not all at once, but by degrees. There is always a mystery in our earliest beginnings. In his ode *"Intimations of Immortality From Recollections of Early Childhood,"* the poet William Wordsworth wrote along similar lines over a century ago:

Our birth is but a sleep and a forgetting
The soul that rises with us, like a star
Hath had elsewhere its setting,
And cometh from afar,
Not in entire forgetfulness,
And not in utter nakedness,
But trailing clouds of glory do we come,
From God, who is our home,
Heaven lies about us in our infancy!

Those first affections,
Those shadowy recollections,
Which, be they what they may,
Are yet the fountain-light of all our day,
Are yet a master-light of all our seeing.

As author John Updike remarked in a 1991 essay, "The poet puts forward a considerably developed metaphysical explanation for the incomparable vividness and mysterious power of our first impressions."

There is a bit of the poet in each of us, and it has its beginnings in that fantastic, never-ending world in which we found ourselves during our first three years. That was a different world, but our guardians were there. Our first words are our appeals to the mother of all and our first eternal father was our own father of course. It was our only world; and it was only there for us. We spent many forevers playing Adam or Eve.

Leaving Eden: From the Garden to Our Own Backyard

Although this information about maturation of the brain has been available for some time, there has been little discussion as to how a constantly changing mental environment is actually experienced by a growing child. It would seem obvious that if the system with which we are thinking is growing more complex every day, our thinking will be growing more complex along with it.

We are all familiar with the concept of infant learning, but we cannot hope to recall the experience of thinking with a brain that was changing so dramatically from month to month for three years after our birth. It is a very long way from the relatively simple mentality of the un-myelinated, un-arborated infant brain to a fully developed adult consciousness that can read books and understand words. If the human brain requires nearly four years to become advanced beyond any other on this planet, from conception to maturity, it provokes speculation as to our earlier mental stages. The workings of an infant brain that is growing more complex by the month differ greatly from a brain which is mature in several fundamental ways.

First, even as recall begins to function, any earliest memories would have to be of a simpler, more universal, nature. Each day we are adding to those connections, so each day things grow a little more sophisticated. As toddlers, then, we must be experiencing an evolving, nearly improvisational, consciousness as we upgrade our awareness day by day. It would be like switching on a computer with the most basic operating system possible and then adding chips daily while revising and improving the systems architecture at the same time. The operating system would have to evolve to match the growing complexity of the circuitry.

A good analogy is the language we speak each day. No matter where we live, we know that our native language has its roots in earlier tongues. Ultimately, this all regresses back to whatever the original human languages were, and we know that they did exist. An American who spoke some German and studied Latin in college might know the origins of half his English vocabulary, but he would be lost in the proto-Indo-Aryan, the source of them all. Likewise, our earliest personal memories are hidden in simpler patterns that a later mind can neither identify nor relate to.

Second, since additional dendrites grow out of the same neurons for years, very early memories will be generalized even further. No matter how memory is made, this rule still applies. If memory is created by a pattern of electrochemical values, it would self-modify as the physical structure underlying it changed. If it is a quantum chaos pattern utilizing neuronal networks, it would be altered by the increasing complexity of the growing network itself.

Third, it stands to reason that those more recently evolved modifications to neural structure would be the last to mature, since they would have necessarily arisen from earlier versions as later improvements. As a result, the last areas to mature are in the sophisticated prefrontal areas which maintain some of their plasticity all the way to adult physical maturity.

The way the brain matures is reflected in levels of consciousness that we employ for common tasks. In 1991, neurologist Larry Squires, working under Dr. Marcus Raichle at Washington University in St. Louis, used a positron emission tomography (PET) scanner to determine the order in which brain structures were being used during recall. He asked subjects to match word fragments to a list of words they had been shown and told to remember. To accomplish this task, they not only used their short-term recollection, they also needed to match word stems with likely candidates in memory. A primitive brain structure called the hippocampus seemed to be involved in immediate recall, but when the mind started word matching, the visual cortex lit up as if the subject were literally scanning a list of words in the visual part of the brain.

Finally, when the mind started searching the associative memory there was a "hot spot" in the prefrontal cortex, as if this structure were now monitoring, or even directing, a search through the entire file of verbal memory. The hippocampus is an ancient brain structure, the human visual cortex is far more recent, and the prefrontal cortex has been doing its most sophisticated memory sequencing for less than a hundred thousand years. From instant reaction to reflective recollection, we seem to activate increasingly complex levels of conscious recall, each level represented by a more recently evolved addition to our basic brain structures.

This leads to some provocative suggestions. If human memory is largely unstructured until the young child's prefrontal cortex is mature, we could not develop a sense of time until fairly late in our mental growing-up process. Our more generalized, undeveloped perception would dull the distinct differences between one day and another, while asynchronous recall would eliminate planning. Months could seem to last for years; years could seem like centuries. Until chronology is established, time remains as endless as a river.

Meanwhile, due to simple brain maturation, memories of earlier images and experiences would be self-modifying and self-generalizing day to day. It's hard to form consistent images of a world remembered so differently from month to month during those three years that must have lasted for millennia; there's all the time in the world before we have a sense of time. Until we can force the past into focus, time remains forever relative; not until advanced brain structures are nearly mature can our experience be precisely recalled or even kept as a reference. "Even if you had the same room, your proportions inside that room would be quite different before and after the infantile amnesia period, so that would make even the perception of the context different," notes Kimberly Cuevas, a psychologist at the University of Connecticut who has studied these effects.

When our growing mind was the flexible place, we were the center of the universe. We were kissed that day because we were so lovable, not because mother won fifty dollars with a scratch card. What did we know of lotteries? We were spanked because we were evil. What did we understand about stressed-out adults or premenstrual depression? We were responsible for it all, since we were the center of the only universe we had known since birth. Before that had been eternity, long before this extraordinary place where things kept changing. We went from pure endless oneness directly to this bliss and misery, from heaven to our own imperfect Eden. It had been forever once, in such endless peace, then suddenly we met these great powerful gods and demons alternately blessing us to dry-diaper-heaven, or condemning us to centuries in too-hot-bath-hell. Sometimes it seemed like forever again, alone in the utter desolation of a dark, lonely room; only to be hugged back to paradise in a mother's arms.

All babies are like that, all over the world. Details are merely cultural; infantile reality works the same in every little unmyelinated infant mind. We were all little angels, sent down to earth; we were all in that fabled garden once. Once upon a time, God really did speak to all of us, thundering from on high. Probably about six feet high, but who's to know at eighteen months with a brain only half way through hookup, innocent of good or evil unable to predict what could happen if we show a smart phone to the goldfish by dropping it in the fishbowl.

But finally the images won't change, and the sequencing is clear. Finally we can remember clearly, seeing ourselves in our minds in a past also sequenced for reference. We become reflective, and begin to see our place in the scheme of things. As our brains mature into memory and clear reflective thought, we begin to pick up and retain both personal and cultural detail. Over a nearly endless time it happens. Our gods descended from heaven to be our mothers and fathers, the great saints and demons take off their halos and horns to become our older brothers and sisters, our aunts and uncles. Bears and monsters become dogs and dump trucks as we graduate from the collective unconscious into the present space through a place of fable and mythology given to us with our baby food. Over three years of worldly time we are weaned from the world of our oneness and rewoven into the collective fabric of our own family and culture.

With the arrival of physical mental maturity, we finally come into this world. The tree of our knowledge is finally becoming fully arborated, and the mind is ripe. Cognitive processing warms up. We begin to notice the many differences between here and there; the differences between me, and him, and her; the subtly differing worlds of early playmates. As we bloom into conscious comparative thought, we are separated from eternity for the rest of our life. We are no longer all and forever; we are fast becoming one more curious soul in the here and now.

Still, even as we all come to grips with the grip of time, there is not one of us that does not distantly remember in some general and diffused manner those days when the gods spoke. We remember the love they gave us, the love that we carry at the very base of our knowledge of this world. It was the earliest language we knew, the source code of our sensibilities. Our very earliest memories start with our parents and their natural love. Babies are treasured universally; there is no culture in the world that condones cruelty to infants. If there is one thing we discover in this awful world that almost makes the loss of eternity bearable, it is the love we found here. It is the only ration that we can take with us when we leave the garden because it is so simple, and it becomes the one compass we always use to find our way back again. We know we must find our way back there some day, back to our old eternal home. We can't forget it just because we are discovering mortality. But we do. We all forget our first eternity. We nearly forget the love as well. But somehow we believe that it will all come back some time.

Back when days were months and months were years, we have the answers to why both Jewish patriarchs and Buddhist devas, those re-born in the blessed realm, had such extraordinary life spans. When we were very small, naturally, "there were giants in those days…" as Bible stories and other creation tales tell us. The years before conscious understanding are so different because we experience them so differently. Nearly all mythologies start with a golden age or at least a time when the gods were making sure everything was working right. It is to this earthly plane that we descend, simply by growing up. The nearly forgotten endlessness of it all that we carry with us is the echo of a much earlier life, and we were there.

When we try to think back to those ancient memories, we can almost scent the breeze of timelessness that beckons over that dark threshold. This is the true time warp, the undertow of trying to remember thoughts from another time, other lives so deep and vaguely comprehended, fossils of the past trapped in the very strata of our mind. We can hardly remember how long it was from age three back to age two. From two back to one is much longer as we move into our collective and universal world time. There is more time on the other side of birth than we can ever remember. There is no time so endless - or so deep.

The haunting memories of those earlier times are still there, scattered and generalized through our waking perceptions, still alive in our dreams and our nightmares. This is the personal and universal mind that is ours alone, and depending on how far back we go, shared with all others on this planet. The further we regress, the more general our entire consciousness becomes, the more time slurs, the more oneness there is in all things. The further we come forward, the greater the differentiation into all the specifics of our self in our space.

Only if the mind itself simplifies can we ever re-experience that other universe that has always been there within us. If the final maturation of our human brain forces us to forget that timeless place in order to deal with this time and space, it doesn't matter. We will rediscover it again at the right time, whenever something makes our mind simple again. It happens every time we let go our nets of perception and find our centers, at moments when time stands still. In terror and in ecstasy the overburdened brain slips time for the moment. Then we can know things that we cannot express or even think about.

It happens every time we undergo an experience so powerful that it blankets our waking consciousness, forcing us into momentary timelessness. It can happen temporarily, but only momentarily, and it keeps us aware that there is some place beyond time. It happens with eternal finality at death, the one and only experience that can actually loose us from the grip of time, and make us timeless before we die.

5

Strangers In Paradise:

The Evolution of Chronology

"The past no longer exists. The future is nowhere to be found. And how can the present move from place to place?
— Nagarjuna

Ever since Darwin it has been accepted that survival often requires novel adaptations to new environments. The latest evolutionary surge in brain development started about sixty million years ago when some daredevil mammals got tired of being chased up trees and decided to stay there. The first primates were already the smartest creatures on earth, and once they got arboreal they never looked back.

As for Darwin himself, his greatest interest after evolution was insects, which come in such variety that he once joked that God must have had an unusual fondness for beetles since He made so many of them. Regard the caterpillar for instance. It hatches, eats leaves, spins a cocoon, and turns itself into a moth if it is lucky. Its neurons are less complex than those in humans, and each may have only a few hundred interconnections. But still, it has 350,000 of them. It needs every one just to operate the two hundred muscles it uses to chew a leaf and that's just for starters.

The compact computational power of a caterpillar brain is beyond any commercial technology we have yet devised. Automated assembly lines with a squadron of robots can assemble automobiles with a dozen digital processors, but the very idea of 350,000 microscopic processors packed into a space that small is awesome. When our domestic robots get that complex, they will probably be able to crawl about and transform themselves into a flying machine too, as any caterpillar can. At this very basic level of mass and complexity, there's more than enough brain power to manage the caterpillar. Unfortunately, it's all used up just operating the insect, effectively preventing insects from building cars for anyone else. When assessing intelligence, the best test we have for conscious memory is learned alternatives, and insects learn practically nothing. The longest memory span observed for an insect so far is roughly fifteen minutes for the scarab, or dung beetle. It actually remembers to feed its young.

We can employ operant conditioning and get responses back from flatworms, but reacting to chemical states is not learning. After pioneering his theories of insect social activity in his seminal work *Sociobiology*, E.O. Wilson began investigating molecular genetics in his later career. In an informal talk at Harvard twenty years after his 1991 Divinity School address, he confessed that he had started to doubt insects could be social. "I began to see how altering something as simple as a single allyl group, such as "hovering near the nest" could, over many generations, proliferate into activities that appeared to imitate social behaviors," he concluded. In fact, if they had the minds to appreciate it, insects could be ideal Zen monks: always in the now, and always in the flow, nothing ever expected because nothing ever happens more than once. A brain that cannot recall a past cannot predict anything. No crisis, no surprises; just processes. So it goes ... forever.

To make any use of memory, it must be retrieved from complex stored patterns, and most creatures simply have neither the capacity to store much peripheral detail nor the ability to sequence mental images into a conscious chronology. Without a true chronological consciousness, animals cannot make any but the simplest plans based on the past. Even chimpanzees, the smartest of the non-human primates, have never made long range plans; the immediate future is all they have in focus. If they could recall even a few years in sequence, they would have noticed the progression of seed to fruit and planted gardens long ago.

Self-awareness is also limited by memory, as we must be aware of the many details that differentiate us from each other. We know ourselves only to the extent that we recall and include those unique experiences that affected us in the ways that form individual personality. Conscious and unconscious memories underlie our likes, our dislikes, our hopes, and our fears. As any sense of self is dependent upon the detail and subtlety of recall, the better memory we have, the more self-conscious we will be. Animals, for all their variegated plumage and behavior, are remarkably similar to each other. All higher mammals exhibit personalities and levels of intelligence, but if dogs had character traits as complex as those of humans, they wouldn't be using their noses to greet each other. Reptiles are so lacking in observable personality that their manner is truly reptilian.

And yet basic communications are possible, emotions are communicated and understood, and one can actually embarrass gorillas and the other great apes, although it is unadvisable. Bigger brains do more than swell heads; they enable the development of complex personal and social structures. In comparison, an insect has no hopes, no bias, no conscious predispositions at all. It never blames, never criticizes, and never complains. There is no self, no self-consciousness, no memory and no meaning. It means more to an observer than it can to itself; its parts are busy operating at full capacity just getting the job done with a mere cubic centimeter of brain matter. There is no recall. There is no time for recall. Without recall, there is no time, no beginnings, and no endings.

The silkworm mechanically pulps some mulberry leaf. A bird overhead sees the silkworm, and recalls a meal. A human notices the bird; remembering and predicting what birds will do to silkworms, shoos the bird away. The silkworm mechanically pulps some more mulberry leaf, a living fiber manufacturing plant with no time for silly things. No time at all. It pulps some more mulberry leaf. No time like the present; no memory of the past, no hope for a future. The silkworm munches on.

Although overall increase in brain mass provides room for better memory and a more refined consciousness, it should be stressed that evolution has never equated sheer bulk with intrinsic value. If that were the case, we would all be under the rule of blue whales. Elephants have much larger brains than humans and they are still using their noses for hoses and working for peanuts. In terms of species, from an evolutionary standpoint, whales are closely related to seals. A very big seal isn't more complex than a small one; just more of. Likewise, there is a lot of whale to operate and everything is more-of including the mass of the brain. Much-more-of, but not better-than. Whale learning has been observed, and it seems to be at seal level, the aquatic equivalent of a smart dog.

This is enough recall for a sperm whale to dive down to where it remembers the giant squids were. The squid, with less memory than a paper clip, never expected anything in its life, far less a large whale in the way. No matter how many pounds of neurons a giant squid was born with, if they're squid neurons it is going to be squid smart and no more. Squid nerves are like cables, so big they're nearly visible. Compared to that level of simple consciousness, even fish are savants. In this world, it seems, any species that can't remember will end up dinner for the rest. It was our last upgrade, the development of sequential recall and projection, which finally lifted us out of the present tense, our mental ascension in the evolutionary development of human consciousness.

Monkeys, Mutations, and Mindpower

How environmental pressures can affect brain growth in a primate was investigated by anthropologist Karen Milton, of the University of California at Berkeley, in a study of two kinds of apes. Apes have a diet centering on fruits and leaves. Fruits provide energy, but are low in vitamins. Leaves are high in vitamins, but require a long digestive tract to extract them. As a result, an ape with a shorter digestive tract must eat more fruit to make up for its inability to thoroughly process leaves. Since most trees bear fruit for only a part of the year, even in equatorial climates, this means that a fruit-eating monkey will need a more complex feeding strategy in order to visit as many fruiting trees as possible in a given period of time. On the other hand, it doesn't take a lot of brainpower to find leaves in the trees if that's our main diet.

Spider monkeys and howler monkeys are about the same size, but spider monkeys are fruit hunters while howlers are leaf munchers. Although both apes weigh about the same, the spider monkeys are carrying around brains nearly twice as large as those of the howlers. The urge to avoid being eaten may have driven us into the trees; but once we went arboreal, locating something to eat, such as a tree in fruit, was the next environmental pressure. One of our most distinctive evolutionary steps was the rapid development of the forebrain's specialized ability to recall and redirect very complex search sequences, clocked and refined by the rapidly developing cerebellum. A fruit eating ape needs to remember and return months later, a task that favors recall and memory.

Under laboratory conditions, when a monkey starts to learn a task, most of the brain activity takes place in the parts directly linked to the event: the visual and motor control areas. When the activity is repeated, however, the forebrain becomes the most active area. Once the event is recorded, the primate forebrain seems to redirect unconscious, pre-sequenced routines that are far more complex than in any other species. As noted in a previous chapter, human subjects utilize a similar memory search directed from the forebrain when retrieving verbal information. In fact, this ability may have developed initially from the need to instantly recall, modify, and repeat a sequence of muscle patterns. This is also exactly what happens if we want to take a flying leap to a swaying branch in the treetops. If we leap to where it is, we miss. If we leap to where it will be, we're alive. What started as an unconscious reflex evolved into a tool for all seasons, making even more use of the densely packed cerebellum.

By the time the apes had their aerial acts perfected, they were using the most complex predictive sequences on the planet. Since they hadn't grown any new organs, there were few changes in the basic operational parts of the brain stem that run the body; instead it was the cerebellum and finely tuned higher brain areas which evolved. Sequential pattern comparison in color and three dimensions requires a lot of complex memory storage. In response, our visual and discriminatory areas added large amounts of new mass and specialization. It is often said that the brain of the porpoise is as complex as that of man, and it weighs more, but its complexities are associated more with hearing than with sight. Primate brains are more visually oriented - monkey see, monkey do. Perhaps porpoises dream in multi-channel color sound. They never say.

It took all of evolutionary history to reach the mass and complexity of the original primate brain forty million years ago. From that time to the present, there have been dozens of diversions from the original line. Some adapted into gorillas, chimpanzees, and orangutans; others, into monkeys, bonobos, and baboons. From ground dwellers to neo-arboreal apes, most learned to get through life with a combination of wits, claws, teeth, and muscle. Some lines opted for miniaturization and speed, becoming the Old World monkeys and remaining in the trees. Our branch of the family gambled everything on brain power.

There was no way to predict that the higher brain centers of this particular hominid would undergo such rapid expansion. Natural law suggests that enough is enough. Most genetic mutations are minor goofs and never reproduce. Still, if it does the job better, evolution tends to run with it; nobody has suggested recalling elephants or giraffes for having big noses or long necks. The extremes are there to define the norm, and there was nothing normal about our unusual brain. It just kept growing. The mass of the brain spiraled upward, doubling its size with billions and billions of new cells. By two million years ago, it had passed 450 cc's and was still growing.

Then something very unusual happened. The SRGAP2 gene that regulates cell growth in the neocortex suddenly duplicated. Brain mass quickly increased. Maternal mortality increased as well, until the female pelvis enlarged to accommodate larger heads. Larger-hipped females became prized since they survived. It also gave women their characteristic hip-swiveling gait, unique among primates, as their legs must slant slightly inward to make up for it. We'd just started making tools when it happened, and it provided some significant improvements. Our homo branch quickly left the australopithecines in the dust as we stopped pounding rocks together and started chipping real axes.

Then it happened again. We gained a third copy of the gene. By a million years ago, our brains had doubled again to 900 cc's, and they kept on growing. It is here where human behavior may have actually supplied the final push, as anthropologist Dietrich Stout suggested in a 2016 *Scientific American* article. Stout was able to demonstrate that the new skills required to fashion good chipped blades require improved three-dimensional visual prediction. One must predict the possible results of each blow, comparing alternatives and increasing the "plasticity" of the brain. Teaching graduate students to flake tools, he demonstrated that the activity actually spurred brain growth itself, acting perhaps as an accelerant at the same time a new level of tool making began to appear.

With our added neural capacity, the finer muscle response we need for gestures and speech also improved as Broca's area, a major speech center, grew larger. Information could now be sorted and shared, spurring the beginnings of language. About 150,000 years ago, we reached our current size, a staggering 1,400 cubic centimeters of mass and complexity. Our brains weigh more than entire monkeys, soaking up a full quarter of our metabolic output like a huge server farm gulping energy day and night. If it took only an extra 150 cc's to give apes the basic ability to predict, recall, and correct, what must have happened to consciousness when mankind strapped on nearly ten times more?

We cannot imagine, because we have been thinking with 1,400cc brains since we started to imagine. We will never know the sort of present-oriented simplicity that characterizes the mind of nearly every other creature. We finally became fully conscious humans when the human prefrontal cortex, accustomed to guiding us unconsciously with its sequenced routines, acquired the incidental ability to consciously sequence our huge memory patterns into clear recollection, abstraction, and prediction.

A promising new area of study, only recently brought into focus by the extensive work of Harvard's Jeremy Schmahmann, is that the cerebellum may have a vastly more important function in the regulation of human consciousness than once believed. Schmahmann demonstrated that in the same way the cerebellum regulates the rate, rhythm, force, and accuracy of movements, so does it regulate the speed, consistency, capacity, and appropriateness of mental or cognitive processes. Although the cerebellum is small compared to the recently enlarged neocortex, it is far more densely packed with neurons, accounting for over fifty billion of them. After decades of studying ataxia at Massachusetts General Hospital, Schmahmann has developed a hypothesis that "just as the motor portions of the cerebellum regulate the speed, capacity, and appropriateness of movements, the posterior cerebellum regulates these same features of thought." As the brain boosted its memory capacity, it was also beefing up and expanding the basic mind-body interconnections that limit and regulate our flow of thought, fine-tuning our reactions and responses.

Developing the Tools for Time

In the primate brain, basic predictive ability is located in the same general area for both apes and humans. A prominent authority in this field, Patricia M. Goldman-Rakic of Yale Medical School, determined that when the prefrontal area is damaged, a monkey's ability to search for a remembered stimulus vanishes. It can no longer hold onto its memory. Such a monkey would be able to use learned patterns to select a tree with fruit in clear view, but it could not remember to return to that tree the next day. Out of sight is literally out of mind.

It has long been known that some individuals suffering strokes in this area appear to lose interest in planning for the future. One researcher who noted this was Swedish neurophysiologist D. H. Ingvar, a pioneer in computer imaging of brain areas during specific mental processes. In 1985, Ingvar wrote: "Lesions or dysfunctions of the frontal or prefrontal cortex give rise to states characterized by 'loss of the future,' with consequent indifference, inactivity, lack of ambition, and inability to foresee the consequences of one's future behavior. It is concluded that the prefrontal cortex is responsible for the temporal organization of behavior and cognition due to its seemingly specific capacity to handle serial information and to extract causal relations from such information."

Analyzing results from tests on humans with damage to these brain areas, Goldman-Rakic obtained similar results. In some manner, the prefrontal cortex had expanded its ability to scan and manipulate memory into sequential projections. Now it was apparently back-loading them constantly into our waking consciousness, allowing us to include an ongoing prediction of possible futures based on information sourced from our vast memory. We take it for granted that we can think backward and forward, but it may be a unique gift. Without this predictive ability, our attention can't seem to get out of the present tense. In another study, neuroscientists P.J. Eslinger and Antonio R. Damasio described the dramatic changes in behavior of an otherwise highly intelligent man whose prefrontal area had to be removed due to a cancerous tumor. He continued to test well, but his daily activity was completely without internal direction.

"EVR" (the patient's initials) was not spontaneously motivated for action. As he awoke, there was no internal, automatic program ready to propel him into the routine daily activities of self-care and feeding, let alone those of traveling to a job and discharging the assignments of a given day. If these goals were presented externally and repeatedly, they triggered the expected actions. But when external recall mechanisms provided by relatives and friends failed, or when the environment failed to challenge him with situations that demanded a response, he resumed his relatively goal-less, unstressed existence."

"The prefrontal cortex can play this role," Goldman-Rakic had observed, "because of its elemental capacity to access and hold 'on line' information relevant to the task at hand. It seems possible that many integrated higher-order functions including language, concept formation, and planning for the future may be built on this functional element." Given sufficient memory capacity, the prefrontal cortex can "access and process information derived from present events and/or long term stores to guide a response over the period of seconds, minutes, and possibly hours required to fulfill the command." The brain's pattern sequencer had been located.

The most striking feature of this type of brain activity is that it provides the outline for what could be characterized as an ongoing neurological "slide show". A typical cruise missile is guided to its destination by moment-to-moment comparison of its GPS progress with a series of internal maps. It seems that our specialized prefrontal cortex acquired the ability to guide us forward or backward in time in a similar manner, by sequencing alternative patterns synthesized from memory. The crucial difference is that in a modern human, the search-and-predict activity has become an integral part of our waking consciousness.

This area of the brain is very recently evolved and would naturally mature after birth. Jean Piaget, an iconic figure in child psychology, described the stage in neural development at which a child watching a toy train enter a tunnel instinctively glances forward to await its emergence from the other end. Before that point, as soon as the train is out of sight, it's out of mind. Here and gone. The toy train's reappearance seconds later is unexpected and surprising. Another train? The mental train of thought had derailed back there when the actual train disappeared from view. This ability to predict was called "conservation" by Piaget, and it appears by degrees. By the time we are three we are aware of the passage of time, and we can finally wonder about tomorrow.

The childlike perspective of constant novelty in the world is unavoidable if we can't store and re-sequence our memory. The prefrontal cortex is the last to mature, so we must slide into our conscious chronology past the age of speech. By the age of four, however, we are experiencing time in our typical three-dimensional framework. Gradually we learn to take such a world for granted, transferring perceptions moment by moment back into memory, creating the illusion that time is moving forward. This may be how we perceive it, but it doesn't mean that time in fact moves in any direction at all. A more important insight is that the three great metaphysical questions could only be posed by a mind that sensed the passage of time, a gift that may be reserved for us alone. This leaves us with an obvious question. Do we create our own time?

The brilliant mathematician Norbert Weiner, whose concepts provided basic foundations for computer control, automation and artificial intelligence, posed a similar question in his seminal work *Cybernetics* in 1948. Weiner reasoned that if time were suddenly to shift into reverse, with planets circling backwards in their orbits, a space traveler arriving on the scene would detect no difference at the planetary level. Time might well run in both directions. However, it would be impossible for forward-time people to perceive a backward-time universe. For one thing, any stars going backwards in time would be drawing in light, not pouring it out. Weiner pointed out that it would be impossible to see such stars, given the way human eyes are made. Communication would be likewise impossible, since the conclusions would appear first, only to disassemble into totally meaningless parts as time receded. He concluded that the only sure thing we could tell about any universe we observed was that it obeyed the same laws of thermodynamics that we do. With the discovery of the vital role of the prefrontal cortex in the sequential arrangement and projection of time-tagged events, another obvious question emerges. If there is going to be a future sequence, why doesn't it simply appear to be our past in reverse? How do we create all the possibilities that appear? There must be a way to create new scenarios without adding anything from outside.

The answer is that if we are to transform our memories into future non-existent states, we must use abstractions to do it. All images of the future are formed this way, using abstractions to create new variations on real images from our experiential memory bank. But where do abstracts come from? Is it possible that pattern sequencing, in humans, may coincidentally serve as an abstraction engine, the ongoing source of the generalities we use to create our future projections?

In performing a chronological scan through patterns, we would be using the clocking capacity of the cerebellum to sort through many patterns, some containing a repeated aspect. Visual abstracts such as "red" might be the result of this scanning activity. Scanning backwards through images which contained first a red bird, then a red flower, and finally a red sunset, the "red" would register three times in a row, more perhaps than other parts of the pattern. If this happens enough times, a "red" sub-pattern itself, as a part of many other patterns, could self-generate, like an echo, creating a resonance leading to a new and independent pattern synthesized from this internal activity all by itself.

It must have created some extraordinary changes in the populations of early humans where this began to occur. *Homo Sapiens'* increase in brain capacity was spread over nearly two million years, but it appears it might have taken only a few small mutations to bring the image-sequencing activities of the prefrontal cortex into waking consciousness. As the largest and most detailed mental patterns on earth were shuffled in a regular and ongoing manner, virtual information began to appear in the mind which was never sensed in the world outside. Networks grew in response to new patterns, and extended them into further synthetic patterns. Even as we filled our memories with a past that happened, we were assembling the elements for conscious abstractions, the crucial process leading to the prediction of any future that might be.

Many religions and philosophies contain hints that this may be the case. The Buddhist saying that the world we see is not a picture but a mirror speaks directly to this paradox of originality locked into the self-generated repetition of our own reflection in everything we perceive. Our final evolution was this transformation of time, as we learned to synthesize information derived neither from genetics, nor actual experience, but through the comparison of personal patterns sequentially juxtaposed in memory. As we project the image sequence forward, we find ourselves in what we call our imaginary future. We can reverse direction and think, "if only I had ..." We can reconsider, and in such reflection, we can be confirmed, or we may regret and we can learn. The word pagination refers to a sequential arrangement of pages; humans reflect and plan using imagination, sequentially arranging, altering, and projecting images from memory.

Automatic ongoing pattern sequencing can't occur in a computer environment because computer memories are reactive rather than active. Neurons are not chips, they are cells. Neither is the brain an immense computer lining up pictures like slides and peering through them. Each cell is alive and pulsing away, night and day. Most of this random muttering is too quiet for us to perceive consciously, but there is always a mental background hum. This continual chatter, a backdrop of constant activity, characterizes the mind at rest, the constant flickering of billions of energy patterns as our cellular chorus carries on the ancient tradition of mindless mental exercise, blending our incidental history with the present moment into the patterns we call thought and the experience we call life.

Waking Up in Africa

It didn't happen overnight. Over many generations mankind grew steadily more mindful. With climate changes, African hominids began to move out onto the savannas. Trees were scarcer and the heat was intense; but we'd learned that by standing up, we could remain vigilant, run faster, and reduce the heat burden. We shed our insulating hair and developed sweat cooling as we became fully bipedal — a balancing act any bird can do but which we mastered rather late — and soon developed varicose veins, lower back pain, and pot bellies as everything sagged downward. The rewards, however, were great. Bipedal behavior offered more than a better view. Now two hands were free to carry weapons, tools, food, and babies. Just as important, the improved cooling system was available for the rapidly growing brain.

The erect posture also tilted the neck up and allowed our skulls to bulge outward in all directions, becoming nearly a hemisphere with a face. The larynx gradually descended, and human speech finally became possible. The first human with an opportunity to be the Adam of our line appeared in a small group of early humans about 180,000 BCE. From this point on, more and more share a distinctive "Y" chromosome, finally including all modern human males. We also found Eve. She was short, African, and the mother of all the mothers of us all. She appeared about the same time, and one of her ancestors had already met a descendant of Adam with the singular chromosome. From that point on we are the same species.

Like our own years from birth to three, the first hundred thousand years of human existence are all but forgotten. We can barely locate them. The oldest human skulls with a modern configuration, from about 150,000 years ago, were unearthed in Zambia at a site called Broken Hill. Referred to as the Cro-Magnon, by 90,000 BCE some of these early humans had moved up country and were colonizing areas in the Levant near the Tigris and Euphrates. They soon discovered the neighbors, some stocky and guttural Neanderthals that had migrated south as the melting glaciers opened new opportunities. The Neanderthals had been enjoying their own evolution, adapting to the more northerly climates for nearly 200,000 years.

Ice age winters were brutal, even for our sturdy human cousins, and these southern Neanderthals thrived. For nearly 50,000 years the species co-existed without showing any distinctive differences in behavior. "Both peoples were living in the same way, hunting the same prey, burying their dead in the same manner," observed Baruch Arensburg, a paleoanthropologist at Tel Aviv University. Then something happened. Suddenly, new technologies began to appear. Better blades, tools, and crafts start showing up - but only in the caves of the Cro-Magnons. Nothing new was happening at Neanderthal campsites except that they were starting to disappear. Soon they were gone.

"I think there was a mutation in the brains of a group of anatomically modern humans living either in Africa or the Middle East," says Richard Klein, an anthropologist at Stanford University. "Somehow new neurological connections let them behave in a modern way. Maybe it permitted fully articulate speech, so they could pass on information more efficiently."

What had actually happened was more profound than speech, and just as unexpected. Ingvar had identified the unusual ability of the human forebrain to create future scenarios. "Evidence is that the frontal/prefrontal cortex handles the temporal organization of behavior and cognition, and that the same structures house the action programs or plans for future behavior or cognition. As these programs can be retained and recalled, they might be termed 'memories of the future.' It is suggested they form the basis for anticipation and expectation as well as for the short term planning of a goal directed behavioral repertoire. This repertoire for future use is based on experiences of past events and the awareness of a Now situation and is constantly rehearsed and optimized."

Was this the final mutation that created modern man? If this is the case, it helps to explain what followed as the Cro-Magnons took over. They were quickly learning to consciously examine their past and imagine a future. More and more of them discovered they could make detailed plans and predictions and remember the whole in their minds while remaining attentive to the present. These imaginary scenarios were not only consciously monitored, they could be consciously remembered. This was much more important than better tools for hunting. We finally had the tools for imagination and abstract reasoning. As a species, we were waking up to a brave new chronological world and it was better than we could ever have expected. It was also cataclysmic for the rest of the planet.

We soon learned the value of practice, planning, and strategy. After survival, procreation is our strongest drive. Nobody made dates before we could plan, and the smart ones soon discovered that better brains were great for mating games. Males can impregnate many females in a year, making females the scarcer sexual resource; and now, for the first time, males could match wits as well as muscles. For the first time, females could select mates with the expectation of offspring and a basic social life together. Together they would travel a new road, working toward a future they could now share in ways they never could before.

Just as each child grows from self-centered to social context, so the world around us was becoming objectified, examined, identified, and structured. Wherever we traveled, we defined our world. We named the beasts, the birds, and the fish. We named rivers, mountains, and plains. In the darkness of night, huddled with the young, we named the demons we all fear; and with gratitude, we greeted the sunrise with praise to Gods named and nameless. With a past and a future in focus, we could dare to be great, intend to do good, and hope for the best. We could also plan to do evil and try to deceive.

Pandora's box had opened inward, and we fell into time as all the miseries of the modern mind flooded into our future, now speckled with doubts and apprehensions, suspicions and fears. Were it not for our hopes and dreams, the bright side of imagination, our newly discovered self consciousness could have driven us mad. We were no longer in harmony with nature but dispersed, as a species, into individual, isolated, time-bound self-hoods. We couldn't be here now and be observing it at the same time. But we were, and we can't stop, even if something in us wants to return to that timeless time when we were part of it all. But it's too late. We became the only ones on the outside, separated and apart, driven out of Eden by our new understanding. The great, great grandchildren of Eve remembered their past and imagined forward, transforming their future. We now had patience to hunt for hours or even days based on future memories and planned expectations. The world was our happy hunting ground, and we headed off in all directions, overtaking and overrunning everything in our way. In only thirty thousand years, we populated the planet.

Those heading east along the coast didn't stop when they got to China. They crossed the land bridge to Alaska and went all the way to Chile in three waves. Paleontologists think the sudden extinction of large native land mammals in the New World about ten thousand years ago was the work of the last human hunting bands crossing the land bridge from Asia. Since the big game had few natural enemies, groups of early hunters feasted their way from Alaska to Peru just by waiting for ground sloths the size of Volkswagens to come home and ambushing them with rocks and spears. We had all the time in the world. Hunting was taxing and dangerous, it was easier to wait around and catch the big ones that never expected us, and the local heavyweights were soon hunted into total extinction. Most ended up as memorable meals for our thoughtful, if thoughtless, ancestors. The earliest hunters must have had an ongoing barbecue.

Although the actual event has not been dated precisely, scientists agree that gene changes, such as the doubling or tripling of SRGAP2, could have had dramatic effects within even one or two generations. The evolutionary change that made chronology conscious seems, as Klein noted, to have appeared fairly suddenly in the Middle East less than 75,000 years ago. From that point onward, the newly thoughtful Cro-Magnons swept northward, some heading for Europe, the others across Siberia. A large group headed back to Africa and, after replacing their slower cousins, started the long march along the coast to Asia. There were obstacles, including a major volcanic eruption in South Asia that slowed things down, but our new update, referred to as *Homo Sapiens Sapiens*, soon met and absorbed the Asian Denisovans and the remnants of *homo erectus*. Some even made it to Australia and ate everything bigger than a kangaroo in about twenty thousand years. This might also explain the rapid disappearance and extinction of the Neanderthals only thirty thousand years ago, the last surviving human variant to share the planet with us.

Crimes Against Humanity: The Cain and Abel Story

Neanderthals were humans, but not like us. Recent studies suggest that the two species underwent a parallel brain expansion later in their evolution. The Neanderthals had more massive bones, stronger muscles, a brain large enough for basic human intelligence, and it seems they had been developing simultaneously. Geneticist John Blangero of the Texas Biomedical Research Institute combined Neanderthal genetic sequences and MRI data to reach conclusions that their brains had a smaller Broca's area for language processing as well as less overall connectivity. "Neanderthals were almost certainly less cognitively adept," he asserted at a presentation of his findings in 2014 before the American Association of Physical Anthropology. "I'm willing to bet on that one."

We know that both species lived, at one point in time, in the same parts of Europe and Spain. In 1991, a Neanderthal skeleton thirty-six thousand years old was found at St. Caesir, north of Bordeaux in France. This prompted Christopher P. Stringer, a paleontologist at the Natural History Museum in London, to declare that this discovery "demonstrates that modern humans and Neanderthals must have coexisted for several thousand years." Richard B. Klein, a Stanford anthropologist, agreed. "Even allowing for some error, humans and Neanderthals were too close together in time to allow one to evolve into the other." In fact, once they encountered the updated Cro Magnons, they lasted less than five thousand years.

The Neanderthals had reached the end point of their evolution. It was the Cro-Magnons who had mutated in Mesopotamia, and they were destined to make short work of their newly handicapped neighbors. As the Neanderthals were settled in Europe first, this might explain the rapid demise of our closest genetic cousins. The lack of preserved Neanderthal brain tissue makes any conclusions about brain structure speculative, but it is likely their awareness of chronological time never progressed much beyond a modern human three or four-year-old. The base of the Neanderthal cranium, moreover, is flatter than ours and unflexed. Jeffrey Laitman, anthropologist and anatomist at the Mt. Sinai Medical School, believes this indicates the higher larynx of a non-verbal vocal tract, explaining a smaller broca's area. Neanderthals probably could not speak excepting in a guttural grunts, chuckles, murmurs, and cries.

They were quite human, however. A number of excavated sites suggest Neanderthals, like other hominids, were communal in nature and hunted a variety of large animals. Judging from advanced arthritis found in the joints of one skeleton, it seems they even cared for crippled or unproductive members of the group. Some suggest the discovery of bear skulls with unusual markings indicates a primitive religion while others, noting flower petals and pollen in ancient burial sites, speculate about the possibility of Neanderthal funeral rites.

However, without a conscious chronology this is unlikely. Possessing basic human intelligence, the Neanderthals would be the cleverest creatures in the woods. Still, their world would remain almost entirely present-oriented, without strategy or analysis. Agile and aware, dim-witted but industrious, they spent their time foraging, digging roots, grabbing rabbits, or mounting ad-hoc hunting parties for larger game. Their campsites also revealed other aspects of their lives suggesting a largely day-to-day response to their world.

For one thing, males and females seem to have lived at a distance from each other and even eaten different diets. No small animal bones are found near the fire sites. It appears males only showed up when bones requiring heavy hammering or burning were brought back to the cave sites. Archeologist Lewis Binford points out these are hardly civilized table manners. "This looks like we've got a situation in which females are essentially taking care of themselves much of the time. Fully modern man obtains food and brings it back. Then it's prepared and eaten by females. I don't think Neanderthals did that."

Furthermore, they were terrible planners. Every spring, rivers in French Neanderthal-land teemed with salmon, and yet there are virtually no fish bones in the Neanderthal caves. "They're not bringing the fish home, putting it in storage, and eating it out of storage," Binford continued. " Modern man plans months ahead of time; they move to places weeks before the salmon run. This all leaves a distinctive archeological record." There is no indication that the Neanderthals were lazy. They just couldn't plan anything past a couple of days, which severely limited their range. "Neanderthals simply didn't make it in the grasslands," concluded Binford. "To exploit the mobile herds of grass feeders, you have to know their behavior and anticipate it. Neanderthals didn't do that. They only lived where food was continuously accessible."

In the near future and the near past, there is time to toy with a bear skull or make a simple scraper, but we find little evidence of careful craftsmanship or planning. Whether a mentally childlike state is a blessing or a defect is a good question, but a mind deficient in both time and abstraction will never be troubled with metaphysical questions. If the only answer to "Where did we come from?" is "From the cave, this morning," there is no need for a past with a purpose or a future with a plan. With memory out of sequence, abstracts could neither be conceptualized nor articulated. Wisdom would come slowly with age and "old" was anyone over thirty.

Our personal years from birth until three included a Garden of Eden for each of us, but it seems our brother species never got out of the woods. Gradual insight gained through experience would have been internal and inexpressible; wordless inspiration unshared and soon forgotten. Abstract concepts such as good and evil were absent from the unmethodical minds of our northern neighbors. Traditionally, the faithful are certain God created mankind, while agnostics suggest that mankind has always created gods. The question really is "When did we develop an awareness that could conceive of God, and what step in brain development made it possible?" The answer is an abstracting and projecting chronological consciousness, and it seems appropriate that it evolved near a place we call the Holy Land. It was anything but a blessing for any other creature on earth, and especially for the locals. Cro-Magnon skeletons dating back ninety thousand years were found at Qadzeh in Israel. However, they found no Neanderthals skeletons in that area from later than forty thousand years ago. Neural upgrades don't leave fossil or bone remains, but while Cro-Magnon skeletons and artifacts continue on to the present era, the Neanderthals simply stop.

It seems we didn't limit our new mental technology to wiping out the local game. We used it to wipe out the neighbors too, and it took less time than anyone would imagine. In fact, if the St. Caesir remains are the last of them, it could have taken less than five thousand years. They were probably well entrenched when the Cro-Magnons arrived. Living in the temperate forests, European Neanderthals never imagined their most dangerous adversary walked on two legs. There is no question, however, that once the newcomers appeared in Europe things were going to get ugly."I see confrontation," says Ofer Bar-Yosef, an Israeli archeologist at Harvard University. "People who grow up in the Middle East understand that. We don't like each other. We rarely intermarry, and we kill each other whenever we can. I don't think you can prevent competition among societies." The arrival of the Broken Hill Gang was worse than invaders from space. They had the best tools, the deadliest weapons, and a real sense of interior decoration, judging from their cave paintings. They also brought curiosity, conflict, and chaos. They came from another Eden, and they took out the native Neanderthals in no time. The locals had no backup plans. In fact, they had no plans at all.

The young Cro-Magnon males had predicted the foraging female would return to where she'd been yesterday and the day before, searching for grubs, berries, or small game. It wasn't exactly like outsmarting a squid or ambushing wild cows, but it must have been easy to grab a Neanderthal. Like the older children in Piaget's study watching for the toy train to reappear, they waited quietly. She was nibbling grapes when they leapt from hiding, converging on the terrified Neanderthal with weapons they had crafted weeks before in anticipation of future events. Now, in the thrashing present tense, they quickly subdued their grunting prey. They probably raped her; later they might slash and kill. It went on like this for centuries. In less than twenty-five thousand years, after millions of years of sharing the planet with any number of variants, we managed to carry out the brutal annihilation of every other human species on earth.

Wherever we found them, it was the end of the road. The only humans who could imagine a God had been given dominion over the earth, and we seized our promised land. The primitives didn't have a chance. They couldn't say a prayer, and they didn't have a hope. They were our brothers, strong and able, but unable to plant a garden or craft a killing tool. We out-planned, outsmarted, and out-bred them. Bearing in mind how we still treat our human minorities, we probably mistreated them, raped them when we could, killed lots of them, and maybe ate a few as well, perhaps with wild flower garnishes.

Modern consciousness may have been our birthright, but it was also their death warrant. For years, scientists argued whether any inbreeding occurred. Genomic DNA testing has given us the answer. They are still with us, dispersed among the very ones that drove them into extinction, mute testament to an unspeakable crime we can never forget. *Homo Sapiens Sapiens* became the only humans on earth: garrulous, upright, and stiff-necked; cursed with the mark of Cain for the systematic genocide of our last brothers on this planet. It is a curse we suffer to this day as we sacrifice our own in war and religious strife. Whenever we kill for the past or the future, we revive and partake in that fearful ancient legacy of premeditated fratricide — the original sin that only a fully conscious human could appreciate or regret.

Part Three:
The Present

6

Stranded in the Here and Now

Have you wondered how
The time is always now
Whether you're a cow,
Or a Mau Mau,
Or an owl?

— *Michael Bridge*

Our earliest mind knew forever, and our final mind will know it again. This exquisite trailing off into the infinite on both sides of mature consciousness may be a suitable explanation for both beginnings and endings, but what about now? If we can close the doors on our comings and goings for a moment, what do these insights say about the present moment?

The present moment, if we think about it, has been going on for quite some time. It has been going on since we can remember, but our attention is often focused in forward or reverse. The past can't repeat, and the future won't be here until it comes, but they affect our every thought. The ability to deal with the present while scanning the past and considering the future is what makes human consciousness such a unique way of thinking.

Unfortunately, embedded quirks in the system end up creating and confronting us with those three questions we all know by heart. "Where did I come from?" "Why am I here?" and "Where am I going?" In the broader philosophical sense they become "Where did it all begin?", "What's the meaning of it all?", and "Where does it all end up?" Our answers form the foundations of our personal metaphysics, our sense of purpose and reality, and our religious beliefs. Once again, the phrase "God only knows" springs to mind, and it seems that God certainly does.

All holy books, including the Bible, the Torah, the Koran, the Vedas, and the Sutras, go to great length providing mutually exclusive answers to these three questions. These conflicting explanations have traditionally served as the basic disagreements separating Shiite Muslim from Orthodox Jew, Evangelical Christian from Zen monk, and Pope Francis from the Dalai Lama. If there are to be some universally acceptable agreements, they will have to harmonize some rather disparate characters. Each religion has its own answers and philosophical or theological structures to support them; each traces its authority to a divine, or at least infallible, being; and they all disagree. A Christian serves God through Jesus Christ, while a Buddhist practices compassion for sentient beings to avoid suffering. The Hasidim are inspired by the mandates of Moses, MBAs by the mantras of mammon. Belief rules our lives, but there seem to be varying sets of beliefs depending on whom or what we believe in.

The articulation of a personal or human raison d'être, some easily encapsulated wherefrom and whereto of life, is what most prophets and philosophers do for a living. If they ever agreed, we would already have a world religion, and it hasn't happened yet. Answers inevitably mirror the complexity, art, and wisdom of their local culture; each explanation a specific response to some universal human need to come to terms with these annoying mental puzzles. Questions of "being" such as these, which appear to be human specific and not culture-specific, suggest once again that we may be looking at an artifact of the neural system. How else would these queries appear consistently only past a certain level of neurological sophistication in human beings everywhere? Perhaps our elegant answers are just the necessary response to something even more basic that makes the same mischief in every human mind.

But supposing answers were available. Would they be universally accepted? If it is the nature of human consciousness to ask the questions, universal answers would have to harmonize the common aspects of vastly different personal realities. Any set of answers that was too personal could become a theology of one, usually regarded as monomania or madness. As a result, in our global society, most metaphysical structures are culturally, but not personally, specific. Which brings us to the question: are we ready for a new approach, and if so, from what perspective? There seems only one response.

The laws of science transcend culture, suggesting the possibility of a source of credible wisdom embraceable by all. The pervasive integration of science into the belief systems of nearly all world cultures is providing us with a common language with which to frame new levels of legitimate inquiry. Rather than seek divine answers that speak to one group or culture, perhaps we could re-examine the questions themselves in the light of basic neuroscience. From a neurotheological perspective, there is one characteristic about these three questions which stands out immediately. They can only be asked by a creature anchored in a chronologically referenced, abstracting, past-present-future sequential time frame. This doesn't even include humans until about the age of about three, when the prefrontal cortex comes on line; and it excludes the rest of life on earth entirely. For everyone and everything else, things just are or they just aren't. We ourselves started the same way. The idea of anything going anywhere in time didn't really figure into our own personal worlds for the first couple of years at least.

The present moment may be our ongoing experience, but we humans have perfected our sense of sequential chronology to give it purpose. Classical Greek has two words for time, *chairos* and *chronos*. Chairos can be "just in time," "the time I figured it out," or "the time of my life;" a momentary personal experience. From chairos we get "charismatic," and the little god Chairos is often depicted traveling on wheels like a Segway, making sure he gets there when the moment is right. Chronos is sequential time, the time that places us in context with our changing world, the time we perceive as passing, and the time we use when we remember or predict. Once we know chronological time, we know it's all going to be over some day. We are the only creatures on earth who know that we will die, so it's little wonder the Greek god Chronos is depicted as a fearsome ancient Father Time. In India, the Sanskrit *kala*, time, returns as fearsome Kali, the black (*kalo*) goddess garlanded with skulls who gets everyone in the end. Indians name little girls Durga, Lakshmi, Tara, all great goddesses worthy of worship; but only Shiva, timeless Shiva, can get a second date with Kali. With the rest of the world living almost entirely in chairos, it was the last evolutionary step of the human brain that placed us as a species on our chronological escalator from past to future. When we began to sequence images from past presents, we finally discovered future imperfects and found ourselves confronted with anomalies arising from this specific new level of conscious time-keeping. One of these anamolies is the source of the major metaphysical questions that have provoked us since we started thinking this way.

Since we can't arrange time until we're well past the age of two, we can't sequence or retrieve full memories back to the beginning. It has been observed that nearly everybody's "perfect imaginary future" appears to be a world similar to "my childhood with me in charge." What this suggests is that we can't imagine further in front than we can remember behind, leaving us with no clear final prediction because we cannot remember where we started. We can't predict "our final future" in our minds without having "our very beginning" to reflect from. As a result, it's our inability to complete the timeline within our own lifetime that keeps us guessing. We sequence back and forth like a ping pong ball that keeps missing the end of the table on either side.

This is annoying. Since we can't access those first beginnings, we haven't the appropriate mental patterns, specific or general, to predict our final endings. The best we can do, in all cultures, is to imagine and hope for a return to an almost childlike and eternal state. This conclusion is provocative in itself. Is this where our notion of heaven comes from? The moment we are able to sense the passage of time, we never seem to have enough of it.

Making Time For Ourselves

We have become familiar with some fundamental concepts basic to our perception of time. The first has to do with our memory capacity, and the other with sequential recall and projection. In fact, the computer itself was defined originally as a device able to instruct itself from an internal program that could access stored memory. The more spacious the memory, the more complex the instructions and the functions of the computer can be.

As to how these memory patterns are created in the brain, the jury is still out; but it's a fact that we simply cannot do anything in a physical environment without leaving some sort of trace. Total and complete disappearing acts happen only in imaginary places. The simple act of typing throws millions of atoms about, sending molecules of me into orbit, and Chinese plastic into space. Physical events are no less disruptive in a cellular environment. Every time a neuron fires, something complex happens in the physical world. When a barrage of neurotransmitters slams into a few million receptors, they don't hit every one; nothing's perfect. There are traces left behind, both in physical molecular changes and in the electrical conductivity at each of the connections. Both have been observed.

Since the average brain may have seventy or eighty billion neurons, even if we use far less than one percent of its total capacity in reacting to anything from soup to nuts, it still involves at least a million neurons. As each neuron connects with up to 10,000 others, the proliferation of one pulse from a single fully arborated axon leaves definable traces in thousands of other neurons connected to that cell. The number of tiny biochemical changes created the moment we wake up and smell the coffee outnumber the stars in the sky.

In fact, smelling a cup of coffee calls into action a full chorus of cellular choreographies grouped and interpreted through personal experience. Neurobiologist Walter J. Freeman, of the University of California at Berkeley, describes the process: "When an animal or a person sniffs an odorant, molecules of the scent are captured by a few of the immense number of receptor neurons in the nasal passages. Cells that become excited fire pulses through their axons to the olfactory bulb.

"The bulb analyzes each input pattern and then synthesizes its own messages, which it transmits via axons to the olfactory cortex. From there, new signals are sent to many parts of the brain, including the entorhinal cortex where the signals are combined with those from the other senses. The result is a meaning-laden perception, a gestalt, that is unique to each individual. For a dog, the scent of a fox may evoke the memory of food and the expectation of a meal. For a rabbit, the same scent may arouse memories of a chase, and fear of attack."

If we could recreate exactly any of those instantaneous energy networks, those individual fleeting electrochemical tapestries as they occurred, we would not just remember our experience. We would re-play our virtual reality and re-live it exactly, a perfect virtual reboot of everything we once experienced. Recall, in comparison, recreates the image or thought with a very incomplete pattern. It is as if the original perception had created a network of electrochemical trails as the pulses proliferated outward through millions of interconnected neurons. Like the after-image of a brilliant fireworks display, the ghost of the event remains in the innumerable synaptic and intracellular changes created when the electrical energy first came coursing through.

If sub-microscopic domains on a flash drive can provide excellent music quality these days, we might wonder why we can't simply play back the past. The reason may be that although the neurological "afterimage" is imprecise, it may actually serve as the physical foundation for another sort of pattern, a quantum electrochemical tracery totally invisible to us except as a mathematical event. This possibility was revealed in a novel experiment described in 1992 by the late Martin Gutzwiller. A prominent figure in field theory and quantum chaos, Gutzwiller came up with an experiment which allowed a single mathematical electron to "drag a pencil" around an area in a random fashion. As expected, the area became darker and darker as the chaotic trails overlapped and merged over time. Then the experimenters imaged it in "quantum time." To their surprise, it revealed delicate and symmetrical patterns, almost like snowflakes. With such patterns, there arises the possibility of comparative functions, the basis of any computational activity.

Unfortunately, though chaotic activity may reveal such patterns when viewed through quantum mathematical algorithms, we cannot think in quantum time. The brain is alive and thoroughly interconnected; the ongoing process is chaotic in the extreme. Just as a pole hammered into a stream will have a definable effect on the flow of water passing around it, so any physical changes, no matter how small, will affect the neural flow patterns in definable ways. This is why every thought changes the brain a tiny amount. Nothing remains unaffected; everything is always changing and changing itself in the process.

Except for rare neurological or dream events, then, memory is rarely replayed. Eventually, patterns not regularly reinforced by repetition through recall lose their linkages and dissolve in the unconscious. Every instant leaves its own record, it seems, but depending on the intensity of the event and the attention we gave it, the currents of time soon wash the labels off our files. They are lost in the warehouse of forgetfulness, where they eventually compost into the general subconscious image bank we use for imagination and dreams.

We had each moment in focus as it happened, and it was all there once; but unless we replay entire sequences over and over again in our mind, memories fade into the past and are soon lost to conscious recall. Studies conducted by James Krueger at the

University of Tennessee demonstrated that during sleep the brain releases substances called *cytokines* which induce special firing patterns among various neuron groups. Kreuger suspects that since even major connections are not always used every day, this is a way of exercising them at night during sleep, preserving the connections for future use. Nevertheless, all but the most extended and repeated networks eventually lose definition and fade from memory.

This is actually a blessing. If memory remained conscious, we would be constantly distracted. We are actually living in the present, so it is better not to have to deal with too many afterimages on the screen of our conscious perception. Asked about his supposed powers to recall previous lives, the Dalai Lama replied that since we should be attentive to the present, his inability to remember a previous life didn't bother him. "I know people who cannot remember what they did last year," he added with a chuckle. "So not remembering an entire lifetime ago is not such a concern to me."

Chimps and Ethics: Moral Codes from Reflective Feedback?

Projecting even a few minutes into a synthesized future is a great leap forward. We know that chimpanzees make the jump. Those familiar with them know that even short-range basic predictive thinking is excellent for achieving their desired results. Chimps can methodically turn apartments into disaster areas just searching for a little food. When surprised by their trainer in the midst of such shenanigans, they frantically try to point the blame elsewhere before sheepishly showing that they know they misbehaved. They know that "if you mess up the place, then you will be reprimanded."

When the brain evolved projection, it unveiled expectation at the same time. Once we could do "if-then," we acquired "should" and "shouldn't". Ethics and morality had their start when we learned to project and expect predictable rewards and punishments in the future for actions in the present. Once we understand the nature of consequences, we learn the rules. This is why personal family interaction during the first three years is so important. All of our primal lawgivers were our earliest caregivers; another reason why holy law often sounds like a parent speaking to a child. When we can realize that we are indeed going to get punished for intentional misbehaving, whether it's eating the forbidden fruit or the forbidden cookies in the kitchen, we start behaving. Trying to instill this sort of thinking in a one-year-old is clearly not going to work. They won't know tomorrow until they get there. Only when we develop enough cognitive sophistication to remember the difference between right and wrong can we try to do right or intend to do wrong. Sin and guilt all kick in when we start learning to be responsible for our actions; we lose our innocence as soon as we know what to expect. Even the Asian concept of karma won't work until we can form a real intention.

The silkworm mechanically pulps another mulberry leaf in timeless eternity. There is no memory in the asteroid belt, no memory flung about the galaxies. It is all happening now, and only now, and everything that doesn't remember knows that. If we really want the odds, the chances of there being a tomorrow are very good. The chance that any specific past or future exists outside our own minds is infinitesimal. Naturally we have problems with time questions. They concern the past and the future, places peculiar to the human mind and personal to each of us. In fact there is only our past and our future, and we made them both up ourselves in our heads when we weren't looking. They are a little different for each of us, and neither is real. One happened before, and the other hasn't happened yet. It's by far the greatest show on earth with an audience of one.

This leads us to one of the harder concepts to master in the neurotheological perspective: the likelihood that all our past and future states are occurring in our minds at the same time, and that the present moment may be all we can ever agree about in any detail at all. Furthermore, if only humans do prefrontal pattern sequencing, the vast majority of the universe must have no sense of time at all. Only a brain of a certain size with certain structures that have evolved to a certain stage could ever sequence chaotic patterns. We have no way of knowing what any other minds might be like; but without those capabilities, the most vital aspect of our human consciousness wouldn't be there.

Unfortunately, since the pattern collections we use for our memory and projection, our past and future, are individual and personal, we come up against a further corollary. We can never ask "where are we all going?" because we can project only into our own personal future. Who knows where anyone else is going? The information that each of us has collected, with which we judge the present and with which we predict the future, is unique. The innumerable coincidences and sensations that create each waking moment before our eyes happen before our eyes only. As a result, we each have a unique and personal mind, a past that nobody knows, and a future that is ours alone. This perspective leads rapidly to the likelihood, previously stated, that there is no provable time but the present moment. Real scientific facts can only occur when at least two people agree about something in the same time and space. This happens when scientists independently observe the same phenomenon, or by agreement, accept the validity of certain procedures and instruments and obtain similar results.

As it happens, no two scientists, or any other humans, have ever completely agreed about the past or the future. Neither have these two places been located by any form of apparatus yet devised. All our instruments make observations in the present. So do we. Any other point of time would have to be memory or imagination, a personal extrapolation or a regression, a prediction in either direction using personal as well as generalized information. There can never be two people in complete agreement about places which exist only in their minds. Generally, yes, but specifically, no.

There's not much we can do about it. Our ongoing sense of time requires that we take images from our version of a past that no longer exists and project them into a future that has not yet happened. Nobody can vouch for us in either place. The only place we can get real, factual, and agreed-upon information is the present. A moment later, it is already reduced to patterns in memory, modifying past patterns, generalizing and transforming into our future scenarios. If we shut down our pattern sequencing, we lose both the past and the future in that instant. Without them, meaning vanishes, and so does time. Memory is the mother of meaning, our lives become valuable to us by that measure. We can lose track of time in the mind, the mind that makes meaning for each of us, in the only world we will ever know.

Metaphysical Questions, Material Methods

We seem to be nearing a reasonable perspective from which to examine some of the questions traditionally requiring religious interpretation. In this instance, the answers to our "time" questions - "Where do we come from?" and "Where are we going?" - become clear almost before a summary. It's no wonder these questions keep appearing. Our sequencer never shuts off while we're awake, and it's always weaving the future out of the past. We don't start collecting a chronological "past" until we're nearly three, when the major structures of our brain have matured, and we can't change it back. As a computer scientist might say about our inability to focus either our earliest beginnings or our final endings, "It's not a software problem; it's a hardware problem." Once we go chronological, we simply can't comprehend the simpler mental language of an earlier time with a timeless mind.

Fortunately, for all our temporal flexibility, we all share the everlasting present moment, the only "real time" that exists in the synthetic sort of chronology we create. It forms the solid fulcrum from which we reach back to recall and throw forward to project. If we want to project forward far into the future, we have to reach that far back into the past. Our earliest past is in undecipherable patterns we haven't used since the age of two. It just doesn't compute past a few fragmentary images. All we can come up with, if we try, is our "best personal imaginary final future" which, as previously mentioned, inevitably turns out to be a thematic version of our personal childhood, with ourselves, of course, in charge.

As our final future is apparently projected from our deepest past during the present moment, it makes sense that both sides of chronological time progress or regress equally back towards that same moment. Tomorrow is the reflection of yesterday; we predict the future with exactly the same accuracy as we remember the past. Both are as real as those images we remember, transform, and project. Both are nonexistent outside our own personal virtual reality and mean nothing to anyone else. In the final analysis, chronological time turns out to be a mental mirror trick that leaves us, like Alice, standing eternally at the looking-glass with views in both directions. We are only who we are; we are never who we were or who we will be. We are now, and were now, and we always will be now. Yet we all remember another then, and project our eventual return to that timeless oneness, our universal memory of eternity.

This is where it began, of course, and this is also where it will end for us. Our sense of time and space, our complex tapestry of experience and reality: it all comes and it all eventually goes. It seems the problems we have with the "time" questions originate with the idiosyncrasy of memory and projection. Something inside keeps hinting that we did, indeed, all come from some same place, and will return to it again. This is why all religions give us explanations for these basic questions, to comfort and to reassure us.

Those explanations were bound to change over time, but the questions still confront us. We may have answers now, as reasonable as they are direct. So where did we come from? We come from the undefined *chairos*, the timeless early mind, into the specific chronology we acquire as we mature. What are we doing here? We are perceiving it all and fitting it all into the patterns we have created through our own personal activities and experiences, each differently, all of our lives. What is our purpose? To be, or to do, what we love the most of course. Where are we going? Back to the same timeless undifferentiated mind that we came from. The cycle is repeated in each of us. We appear out of our own timelessness; we put it into our personal perspective and temporary definition and then, finally, we must blend back into that same timelessness again

7

Feel Is How We Real
The Meaning of it all

> *"The question of whether the world is nothing but a physical accident or whether there is a plan, this is the main question of every human being. Because the only answer to our suffering would be that there is a purpose in it, that there is a spirit behind it. If these would not exist, our life would be a hopeless business."*
>
> Isaac Bashevis Singer

It's hard to argue the reality of the world we perceive around us. If we do it too often, people think we're neurotic. Since we share the same DNA, at the most basic level our senses are giving us all the same picture. There are some physiological differences, such as color blindness, but they are rare. As a result, we all appear to live in the same world when we are actually all experiencing our own personally created variations. The rarest thing in the world would be the most normal viewpoint of all, right at the top of the bell curve; the perfect average. We are all off-center by varying degrees, experiencing a unique virtual version of the shared reality of humankind. Viewed from this perspective, any description of the nature of the world we live in could represent only one version of a generalized, consensus reality. As such, it would also provide a penetrating insight into the mind of the perceiver. If our world is not actual reality, but a representation we create, we might consider Singer, a Nobel laureate, as being more self-descriptive than perceptive when he made his haunting statement.

Singer, author of *Fiddler on the Roof* and a Nobel laureate, was a gifted writer. His ability to spin tales of insight and wisdom, embodied in the lives of the memorable folk characters he created, is close to the very definition of the artistic mind. At this level of creativity, he would have had an internal world as vivid as any perceptual world existing around him. He died at eighty-seven in 1991, fulfilled and famous for decades, still overshadowed by powerful memories of suffering in a world long gone. The Yiddish word for such troubles is *tsuris*, and there must have been enough of it at one time that he still sensed it daily in his mental landscape despite all the great good fortune and international recognition he eventually enjoyed. He still yearned for a plan, an explanation, a reason for existence, despairing if it were not found.

He was searching for the big answers. This is a common theme among the creative, but absent among the dull. Philosophers search for meaning as consistently as a walrus digs for clams, a sort of instinctual preoccupation. Of course Singer wanted a big plan. That's what all philosophers want. At the top of the bell curve, most people, less artistic and much less philosophical, are searching for big discounts, and they want a deal. If they like it, they buy it. The Zen monk says, "Make me one with everything," so the hot dog man gives him a tofu chili cheese dog with sauerkraut. Is life profound or is it just a joke? Has it a plan or a plot? Are we saved or are we damned? Is it even worth worrying about one way or the other? These eternal inquiries return us to a number of thematic variations on Bertrand Russell's question, "Is there a purpose to the universe?" Of the three basic questions - "Where did it all come from?", "What does it all mean?", and "Where does it all go?" - the middle one seems to come to mind most often. Neural function may well be the basis of chronology and abstract thought, reducing them to the level of operational phenomena, but this doesn't answer the reality question. Is there any real meaning to any of it? Is there any ultimate truth? Once again, insights may lie in the way the system itself seems to work.

Seeing sunsets with signal-to-noise ratios is mechanical consciousness at chip level, the cellular mechanics of our human color vision system. We have witnessed how, over time, both as a species and individually, brain development and maturation allowed human consciousness to arise in an example of dynamic systems development over which we have little control. The recent acquisition of forebrain pattern sequencing capability for both time and abstraction may simply have been the most recent human upgrade, allowing us to model and manipulate our current mental perceptions. All of this is built in. It came with the kit; we can't change it a bit. We hadn't gotten into personal issues yet.

Discussing answers to questions about either personal meaning or ultimate meaning is different territory. We must now venture into personal belief systems, a form of mental application software that can, in fact, be updated by personal experience. As usual, rather than answering the questions directly, we might do better to investigate the system that makes the questions and see if there is a little common sense lurking about in there. We may have an idiosyncratic neural architectural, but the rules of architecture themselves don't change from model to model. Likewise, we all have a different fix on reality, but the system we use doesn't vary from one mind to another.

Tabula Rasa: Making it Personal

Neurologically speaking, we are born into this world with brains as blank as cauliflower. Operationally, we are living and doing a good job of it. but without a clue as to what is going on around us. From a meaning point of view, we are at zero ego, which makes meaning moot. If we are going to perceive meaning, someone has to be doing the perceiving, and we have to develop that someone part first. A human infant takes at least ten months to discover that it is separate from its environment; it takes two more years to really get things straight. Reality isn't what we need at that age; we need security. Without thinking about it, our earliest experiences and environments start to lay the foundations of personality, our unique way of dealing with our world. It happens without effort and without thought, and most of the events are entirely coincidental.

Our very first perceptions start the ball rolling. No matter what sense is picking up the information, it is sent off to the brain for correlation and reaction. The brain, at this point, is quite plastic, still maturing, neurons still branching, axons sprouting new dendrites, pathways pruned day and night as we upgrade to full human consciousness a little past the age of three. The basics are in place early, so for months those neurons have been doing their compare and reset, compare and fire routines.

Each time a pulse sends a neurotransmitter into the synaptic gap between two communicating neurons, it changes the biochemistry of that gap. The pulse is actually relayed by the neurotransmitter molecules, which diffuse across the gap and fuse with the cell membrane of the neighboring neuron. What divine force impels them to their appointed uptake sites? When it comes to the most fundamental questions, answers can be disarmingly simple. They're moving in a fluid environment and they pick up static charges like laundry in a dryer. Nothing is perfect, of course, and a lot of the molecules end up floating about like space debris in orbit. Some make it back where they came from, a process called re-uptake, but some never do. Every time the neuron fires and the routine is repeated, every synapse has more bits floating about in its communications channel.

Altering a synaptic gap with more or less leftover space junk doesn't have a great deal of effect on neuronal activity. Each neuron has thousands of synapses, one at the end of each dendrite. Still, there's more chance of getting a billiard ball in the pocket if you have dozens of them ricocheting around, and the same may hold true in intercellular space. The net effect is a drop in resistance at that synapse, and more pulses get through. These lowered-resistance synapses are created at the moment perception or recollection happens. They are the physical byproducts of momentary energetic networks that are created by the tens of billions throughout the brain at every moment as millions of neurons send their computations across thousands of channels dozens of times each second. Every instantaneous energy network leaves its ghost in the system, reinforcing previous levels of molecular alteration already created by previous events.

Investigating synaptic change caused by repeated stimulation, neurologist Gary Lynch, of the University of California, discovered that repetitive activity by itself can create permanent physical alteration at the synapse level. In his research, Lynch observed that neurons in the hippocampus and cortex would remain more responsive to repeated stimulation for weeks or even months. The phenomenon, called long-term-potentiation (LTP), is believed to play a part in the formation of basic memory. Since any perception similar to an earlier one will utilize some of the pathways created by the first perception, it effectively enlarges the pattern. In fact, it was demonstrated that in a dark room, a small flash of light followed by a bright flash could produce an interesting effect. The subject often had no memory of the first flash. The second flash, barreling down the same channels as the first, had obliterated nearly all the traces of its predecessor.

Simple repetition will increase the size of the neural network as some pulses will now travel further along lowered resistance pathways, contacting a larger population of cells. This simple rule is the underlying basis of our personal sense of what is real. Our initial perceptions may be incidental, but repetitions can be more than coincidental. If it happens again, it starts to leave a trail. We are soon being biased toward the repetitive, and the resulting networks will shape our lives. By the 1990s, scientists were creating "neural net" programs which could actually "learn" in this manner. The software lets multiple connections interact with each other, automatically reinforcing those pathways which get the most traffic. If this appears to mimic the basic Google ranking algorithm, that's correct. As early as 1985, computer scientist John Hopfield had perfected the first neural net software program. It was able to imitate nearly exactly the neural processing of a sea snail. The program not only reacted correctly, it "learned" patterns of attraction and avoidance as it encountered software equivalents of pleasant or unpleasant stimuli.

The sea snail, *Aplysia*, was chosen for the experiment because it has only about two hundred thousand very simple neurons, well known to all, making it a favorite for neuroscientists. Of course, real sea snails do it all under water, and make more sea snails by the sea shore. Synthetic neural net programs have grown in their complexity and applications, but even Watson's mega-computer clouds cannot begin to approximate the millions of living networks that are modified each moment in the human brain. A recent assessment suggests the human brain could easily perform 10 million billion operations per second.

We now know that any single neuron may release two or more different neurotransmitters at a single synapse. Some neurons can switch molecules, as if changing languages, and send different messages at different times depending on what else is happening in the body. At this time, the number of known neurotransmitters has increased to more than eighty, and nobody knows how many will be found. The relics of innumerable previous energy networks established by these uncountable electrochemical changes are subtle indeed, but repeated stimulation allows some to grow, while others wither. Some neurons develop additional uptake receptors and even new synapses much later in life. The neurons in the prefrontal lobes, for instance, are not finally wired until late adolescence, allowing us to intensify or even alter, aspects of our working mental system well into adulthood. This also helps to explain why teenagers everywhere think they can predict better than they can. Soldiers, also, tend to be young. If they could really weigh their possible outcomes, it would be hard to recruit an army.

In the infant, early experiential traffic begins to build both major and minor pathways, predisposing the flow of any later activities. This happens with our very first perceptions, and it never stops. Our earliest fetal memories were deep and timeless but nearly featureless. Shortly after birth our neural patterning begins in earnest. Since neurons fire, rest, and then fire again, an infant can continually enlarge a neural network by simply staring at something for a while. In rural China, in a basket next to a rice paddy, an infant might be encoding a pattern based on the planters; another, next to window in New York City, might do pigeons. We all create patterns based on parents' faces and familiar sounds, on family smells and tastes and touch and everything available. Soaking up our surroundings like a sponge, we furnish our house of thought with whatever is there at the time. Our unique sense of reality is based on this broad unconscious network, the ancient neurologically entrained part of early memory itself. Over time, simple repetition creates idiosyncratic collections of extensively interconnected patterns, whatever is most repeated during those times that our earliest neural nets are being created. If there is a friendly dog in the home, we make early networks based on its smells and behaviors. We become young children who were familiar with dogs, knowing instinctively some of the general signals that distinguish playful from menacing.

These physical and electrochemical networks become far more extensive in the human brain simply because of its greater mass and complexity. Since our body hormones react in accordance with our perceptions, internally as well as externally, the size of the neural net begins to have an effect on our perception itself. The electrochemical energy expressed by the energizing of a growing associative neural network, less resistant and more interconnected over time, will gradually trigger larger and larger hormonal responses. Because of the extraordinary density and size of the human brain, there comes a time when these hormonal reactions to the energizing of those neural networks can become equal to, and occasionally even more intense than, our original reactions to the experience that first encoded them.

Since several major hormones associated with levels of brain stimulation have perceptible effects on internal body systems, we eventually begin to perceive a "feeling" that accompanies some perceptions and not others. It is the body's hormonal reaction to the energizing of those neural networks, an electrochemical echo that is there in varying degrees depending on the relative size of that network. As size is a function mainly of repetition, our emotional spectrum is essentially a function of our familiarity with everything we know.

These hormone-based feelings soon become our basic underlying biochemical checkpoint for the simple safety of past experience. It has happened often before, so it is familiar. We feel the way we feel when things are all right; it's making sense. Actually, it's making microvolts. As repetition begins to provide hormonal clues as to what is safe and what is not, the tendency to remain safe by repeating the familiar builds upon itself as soon as the infant is able to express choice. Getting toddlers to accept variety is sometimes quite a chore once their little minds begin to identify what they like and what they don't.

As our personality is defined by the way we feel about everything we can think of, we become who we are primarily because of those early experiences that influence our basic mental preferences. Our sense of ourselves is an unimaginably complex pattern of continuing hormonal clues that had their beginnings in pure chance and parenting patterns, the earliest repetitions of our infancy and early childhood. In fact, all our feelings arise from our varying hormonal reactions to both external and internal stimulation. The more neural activity we generate in response to our perceptions, the more we will feel excited or aroused about whatever we are perceiving or imagining.

This is unique for each of us, and constantly changing. As neurobiologist Larry Swanson, of the Salk Institute, describes it, "Instead of thinking of nerve circuits as fixed anatomical circuits that always do pretty much the same thing, there's a metabolic or biochemical plasticity, a real chemical dynamic in brain circuits that is probably different to some extent in different people." Even after our innate patterns of personality are set and neurologically reinforced, personal thresholds for triggering fear, pleasure, depression, and delight are always being set and reset by the growth or the waning of neural networks as they react to events, both external and recalled. Whenever repeated experience and thought add further networking, we feel more intensely about it, whatever it is. By the time we have encoded any routine deeply into our memory, we have thought about it and imaged it many times. Each time we did, the network enlarged, and each time we do it again, it will enlarge a little more.

It may sound artificial, but if we want more meaning from anything all we have to do is concentrate on it and it will, to that degree, become more interesting and mentally attractive. For example, if we were to arise each morning for a month and sing the same verse of a song and then go out and stare at one particular tree for five minutes, both the song and the tree would acquire greater meaning for no inherent reason at all beyond the time we had spent with them. Gustave Flaubert noted, "If you want to make anything interesting, you simply have to look at it long enough." Repetition alone, even in adults, expands the associative network ever wider, allowing its effects to eventually even influence deeper layers of consciousness. There are many mental exercises that use this effect very effectively. Repeated prayers or mantras are a good example of thoughtful repetition helping to structure a reality where security is associated not with the daily interruptions of life, but with quiet times of emotional and mental peace. Over time, nearly any form of mental or physical repetition, from meditation and prayer to dance and practice, will in this manner become a source of consistent internal stability and emotional well-being, as comforting as a personal friend. A ballerina acquaintance once remarked that the mental focus on her body required for mastery of her art was probably better than therapy, saving her "buckets of money" in a stressful profession.

As we progress from infancy into early childhood, our personal memory networks grow day by day until there are enough solid associative cross-links to provide an ongoing emotional sense of general familiarity with nearly everything we perceive. Our world becomes real around us, a mirror of our own memories and a memorial to the particular path we have followed. From this perspective, we might reflect again upon the sorts of pain and confusion that must have affected Singer's early life, experiences that caused him to feel that his world was facing such futility. His tales resound with humor and ache with compassion; but he himself was unsure, yearning for the certainty of an elegant plan. As the neural twig is bent, so the arborated mind will grow; it is nearly impossible to underestimate or erase earlier realities. Children who get assurances when they need them usually grow up to be self-assured adults.

Interestingly, the effects of brain activity in the origin of emotional states can be illustrated with a simple graph based on a parabola. In this linear view, we can observe the effects of a larger and larger hormonal signal evoked by either outer sensory stimulation or internal networks. It is interesting to note the manner in which familiarity or unfamiliarity serves as the primary basis of nearly every emotion from ecstasy to terror. The line intersects the sense of "reality," dividing the external and internal sources of an emotional state.

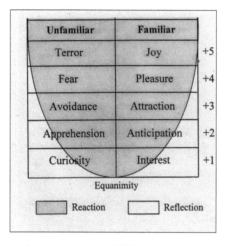

The reason is that we react differently to increasing stimulation depending on whether it is based on the novel or the familiar. If it is familiar, it energizes neural patterns that are already networked and sequenced; if it is not, the brain must speed up immediately to gather more information and determine a timely course of action. Whenever feelings start to arise, we can be certain that neural activity has increased. If the source is coming from outside, we're trying to find a fit for a growing unknown, which could mean danger. If the activity is from triggering a large neural net, the hormonal "echo" assures us that it is welcome, or worrisome, depending on previous experience. The known danger can often be dealt with, but the unknown is always dangerous.

The midpoint of this graph is at the bottom, representing the mind at equilibrium, equanimity, where the mind's reaction to perception of life is not apparent to us as any emotion at all. This is true disinterest, when neither a positive nor a negative feeling is present. Our emotions increase by degrees, familiar on the right and unfamiliar on the left; null to pure joy on one side, null to pure terror on the other. These gradations are, in fact, levels of reaction to greater and greater amounts of brain activity. On one side is reaction to increasing levels of unfamiliar information without reinforcement from established networks; the other side represents the reaction as familiar information activates larger and larger networks already in place. On either side, at the top of the curve, the adrenalin surge will always leave vibrant memories.

One interesting effect will often occur when the environment provides terrifying or ecstatic scenarios. The hormonal overloads creating these feelings always speed up neural activity, creating the sensation of the world slowing down. At the same time, if intense enough, it momentarily blots out nearly all personality specifics. This is the reason that soldiers bond in battle and couples bond in bed. As boundaries dissolve in terror, all men become brothers and in ecstasy all couples become lovers. Of course, sadly, it wears off, but the memory of that momentary sharing is the basic misunderstanding that lies behind both jealous possessiveness and the glorification of battle. We all seek that oneness of place and purpose, but sometimes we have to jettison our personalities to find our partners.

Perception usually overrules reflection, but on some occasions a great amount of mental arousal is clearly an internal, rather than an external, phenomenon. Finding a photograph of one's mother is going to churn up a lot of networks, usually positive, even if nothing else is happening. We smile with happy memories. Likewise, being trapped in a car careening towards a freight train would cause rapidly increasing stimulation from the senses with little familiarity as to how to deal with it. The mind races to sequence and predict, but finds blanks rather than reassurances. Frantic neural activity releases hormones that propel us instantly into feelings we now identify as terror.

Of course, if the person in the car is a seasoned stunt man who chose his dangerous career because of stunted emotions due to an abusive mother, the situations would reverse, with feelings of pleasure and panic switching places. It all depends. Our personal emotional makeup is so very personal and idiosyncratic that even the ancient Romans had a saying, "*De gustibus non disputandum est*" - "There's no accounting for taste". The mechanics of preference are, in fact, no more than a construction of familiarities. Likewise, our responses to the virtual world we perceive are continually biased by the internal mirror of previous perceptions that we collect throughout our life.

As a further illustration of the way in which familiarity, or lack of it, forms the basis of emotional reaction, there are the parallel parables of the book and the "bliffer." In the first instance, we imagine we are walking down a sidewalk in a familiar neighborhood, thinking about nothing in particular. This would correspond to the point at the base of the familiarity parabola, the absence of either positive or negative feelings. We spy a book lying on the sidewalk ahead. This raises us a notch from null to +1, interest. We know what books are, and here's one lying on the sidewalk. The "book" networks energize as we approach to get a better look.

At closer range, we see that it's a spiral bound notebook with what looks like the seal of the college that we attended. This is greater familiarity. The neural circuits energizing with memories relating to our college years increase the hormonal release to the level we associate with pleasant anticipation, +2 on the familiarity scale. As long as it feels good, we will approach. Bending over the notebook, we see that it is indeed from our old college. The rush of conscious and subconscious memory boosts the hormonal level again, this time to +3, and we feel attraction. We are smiling and warming internally. The endocrine release initially relaxes capillary walls, creating a radiator effect, and causes the liver to release blood sugar, improving muscle metabolism. This feels good. More old familiarity patterns log on, increasing the emotional "meaning" of the situation.

As we reach down to pick it up, we are amazed to realize it is one of our own notebooks, one that had been in a box that had been misplaced years ago. Our pleasure increases to +4, pleasure. More adrenalin. More ACTH. Neural firing increases; our capillaries tense with a shiver of expectation. We clutch at the book, bringing it even closer. Now we are really getting excited because maybe this is the long lost notebook with the only record of the address and phone number of a friend we have been trying to find for years.

It *is* the long lost notebook! The number and the address are still there! We have now topped out at +5, pure joy. The entire day seems to have stopped in its tracks as the rush of hormones floods us with what we interpret as our pleasure at finding not only a familiar old notebook, but also the chance of reconnecting with an old friend. If anyone were watching us from a distance, our rapid changes from one emotional state to another without any obvious change in the world around us would seem most peculiar. In fact, there is every reason for us to get excited about something so interconnected with so many of our internal memories. It's the way we separate the familiar from the unfamiliar, ranging from the interesting to the delightful. It is our mechanism of meaning, the way we can use our massive memory to provide a much wider range of emotional clues than any other creature on earth.

On the other hand, there is the bliffer. A bliffer is a quasi-android intelligent pseudo-life-form from the Dwaroid galaxy, one of those going backwards in time relative to our own space-time perception. Bliffers are often given as pets to young zurks because they absorb bad dreams. They are trained during their autogenic cloning to seek out any apprehension or fear. Like zurks, bliffers are metallic and communicate with vapor streams. In other words a bliffer is a harmless semi-living pacifier that fell out of a flying saucer passing between where we can't see and where we can't know. It is also unknown in this universe. There has never been a bliffer on earth before today.

Once again, we are walking down the same sidewalk. This time, we notice a little metallic object lying on the sidewalk ahead. It is unfamiliar, but small, and the increased stimulation we get from directing our attention to it is also limited. It wakens a couple of familiarity networks associated with cell phones or cameras, but there's no match. We feel curious, +1 on the unfamiliarity scale. When it comes to approach/avoidance, the first stage of avoidance is usually a form of approach. Depending on the circumstances and our feeling of security, we approach the object mainly to acquire more information so we can figure it out and fit it into some association. The bliffer is just lying there, so we pick it up to examine it. It "wakes on" and starts to warm up. At this point, we notch up to +2, apprehension. We start to tense internally. Then, as bliffers will do, it starts to hum. To us, it's acting strangely as unexpected visual and tactile stimulation start to occur without any internal correlation. Is it a smart phone with a battery about to flame out? The mind is beginning to feel the effects of increased hormonal release as our emotions rise to +3, avoidance. We put it right back down on the sidewalk.

Now that it is warmed up, the bliffer tries to ask what is happening, but as far as we can tell, it just started to smoke. As the vapors begin to rise out of the device, we back off. This thing is really weird; maybe it's a terrorist bomb about to explode or something. We recall images of car bombs, triggering +4, fear. Sensing that a nearby life form is imagining a nightmare, the bliffer does what it was made to do and slowly slides towards us. That's enough to nearly terrify any unsuspecting human. We turn and run, with the poor bliffer, still trying to do its best to calm us down, skittering along behind us. We're scared silly, +5, the top of the scale.

In both examples, nothing much has really happened. In one case, internal familiarities elevated us to a froth of joyful anticipation while in the second, the lack of those same familiarities had us running scared when no danger actually existed. The book itself had nothing inherently pleasurable about it, and the bliffer was harmless. It all happened in our minds. In fact, if we want to be completely mathematical about it, it could be said that joy is familiar information building up faster than we can integrate it, while terror is unfamiliar information flooding in faster than we can get away from it. To us it is all kinds of real, but it's no more than perception, patterns, and hormones.

Does mental complexity have anything to do with it? There is no reason to expect that those on the slightly more interconnected side of neural structure would be coincidentally blessed with sluggish hormonal systems to compensate for their excitable minds. Like high rpm engines, they always run their endocrine systems at the higher end of the scale, living their emotionally vivid lives no matter what the weather. Race cars run at very high speeds, yet in experienced hands they are as safe as station wagons. The denser the interconnective network, however, the faster the emotions also go up the curve. More pathways mean more electrochemical energy, which means more intense reaction. The over-imaginative can fall in love after one meeting, all excited with hormones and hope. The paranoid can intuit real danger in the arrangement of towels on a rack. Ruth Richards, a researcher at Boston's McLean Hospital, has suggested that the emotionally vivid lives of those suffering from bipolar disorder may explain the number of writers and poets who were known bipolars, including Shelley, Byron, Hemingway, and Woolf. The dull rarely get emotional about anything that isn't obvious and often don't sense danger until it is nearly too late. Each one of us is completely different.

The way that familiarity eventually weaves meaning around us was described by the late neurophysiologist Jonathan B.B. Earle, scholar and researcher at Bradford College in Haverhill, Massachusetts. Earle noted that both drug and meditation induced hallucinations develop in a characteristic three part sequence. In the first stage, the individual is aware of what seems to be a confusing pattern - a visual hallucination, a mantra, a mandala, meaningless images, or a jumble of sounds. After a time, the patterns begin to acquire meaning, and fear gives way to familiarity, acceptance, and a sense of understanding. Finally, the individual feels a sense of actually merging into the pattern, taking part now in a meaningful experience, no longer confusing but deeply reassuring.

This is but a rapid example of exactly what happens to each of us over time as memories and predispositions urge us to repeat and repeat enough to drive their associations deeply into the most basic structures of the brain. All during our lives, this gradual progression is repeated. The graduate student in her first semester at law school is awash in terms and torts, contracts and court procedures, trying to make sense of it all. By the end of three years, the theory and the systems of law have become a complex and comfortable structure with both meaning and purpose. The complexity and confusion of law school transforms in a chronological progression into the comfort of a legal career. After ten years of legal practice, the structures and systems of the lawyer's practice have become an inseparable part of her daily life and thought, now thoroughly integrated into both self image and life direction. Once intimidating, the practice of law is now reassuring.

In a like manner, we are all originally confronted with life as a confusing mass of stimulations and perceptions. Gradually, as we move into our self-repeating repertoire of likes and dislikes, we find ourselves drawn toward the patterns and images most agreeable to those internal patterns we have already woven through our repeated experiences. Finally, over time, as we become more familiar and expert, our own lives begin to provide a more personal and profound sense of meaning. This becomes the basis of our adult sense of self and of our place in the world. This personal self-creation of meaning can be a frustration to our many attempts at communication, but it is vital for any sense of personal validity. We perceive with a consciousness both unique and consistently idiosyncratic about what is real and what is not. If we did not have that sense, we would have no sense of ourselves either; without a personal outlook on life, we would be reduced to herd mentality or ant-like anonymity.

Eastern philosophies seem to understand this paradox somewhat better than those in the West. Returning to the Sanskrit word *maya*, its three translations ordinarily might seem in conflict. Maya means "beauty," and it also means "power." But it means "illusion" as well. *Maya* is the power of our beautiful illusion, our one-person personal virtual reality, to fool us into thinking that anything has inherent reality or meaning beyond our need for that hormonal checkpoint, our search for familiar cues that feel real to us. It is *maya* that makes us imagine that we will be young forever, that power is exhibited by money or intimidation, or that we are here only to be gratified. It is always waiting to trick us into believing that something in the world is awfully important, when in fact everything can be reduced to idiosyncratic perception constantly biased by personal memory.

We see it, but we just don't get it. *Maya* is forgetting that everything we sense is, in a way, only temporarily there and subject to change at any time. *Maya* is thinking that we really know what is going on. It's a nice thought, but it can't be true. We can only know what our own neural networks filter through to us, hopelessly biased and idiosyncratically perceived. Only if that idiosyncrasy were eliminated could we experience the real universe directly. As long as we have the bias of our ego, we can't be objective in any way, not with a brain that operates with neurons and synapses. If we cleaned up all the incidental debris and rewired the nets practically rather than incidentally, we'd wreck those personal patterns, and we'd be back to ant-thought again. We could all perceive the same world only if we could agree, even for a moment, to retrofit everyone with identical brains to see how things really appeared when we were all perceiving with the same mind. This is not going to happen.

The Buddha reflected on this with his well known observation: "There is no Nirvana without Samsara, and there is no Samsara without Nirvana." "Nirvana" traditionally means a state of no bias, no time, no space, and no ego either. "Samsara" is experienced as a life of futile repetition, likened to a fly caught in a vase buzzing around and around without getting anywhere at all. Nirvana is unbiased, untimed, pure perception. The repetitive and circular nature of life that most people experience, repeating again and again due to the tug of personal emotional attraction and avoidance, is Samsara. There is no reason at all to suspect that Siddhartha Gautama Sakya was familiar with neural nets or virtual realities for that matter. As an ex-prince, he was the junior executive dropout of the fifth century BC. His statement, however, speaks to the paradox of reality that we all face. If we approach Nirvana, or experience it, we would be perceiving true reality in its purest form. To do this, we would somehow have to circumvent or eliminate every idiosyncratic associative interconnection in our neural system.

This is all fine and good, but it essentially reduces us to the level of human cauliflower again. Enlightened cauliflower perhaps, but without those cross links, incidental networks, and hormonal clues, we would lose our sense of meaning, our sense of reality, and our sense of self at the same time. To know ultimate meaning, in other words, we couldn't be we. We would lose ourselves in the very process. This might be fine for a last moment of conscious existence, but it doesn't pay the rent. Perfect nothingness may be neat now and then, but it's also without worldly value. It's nice to know it's there, but it's no place to spend a lot of time.

Likewise, whenever we think we have things figured out so that we are finding enormous amounts of meaning or misery in life, we are just over-focusing on ourselves. We're getting a lot of excess activity overworking our personal mental networks, hopelessly out of touch with true nature of occurance. It's hard to know what is actually taking place around us because it's so entirely biased by these very personal filters of preference and familiarity. It's a paradox, but it is true. We can't have one without the other. Without the ground of meaninglessness as our baseline reference, we can never have a real feeling for meaning.

This sounds like saying we will never really find it unless we lose it. It's no coincidence that all spiritual teachers say we will never find ourselves unless we can lose ourselves and reclaim ourselves again. As infants, when we see ourselves in the mirror, we are just part of the picture. Not until we're ten months old do we know that someone is there. This complete lack of self-awareness, temporarily or permanently, lies at the heart of religious fulfillment. Meaning is moot in infancy; we simply are. From that point on, we make ourselves and our sense of meaning at the same time.

Our personalized, homemade feelings define us, they direct us, they make our lives meaningful, and they bring us all the emotional pain and mental suffering we know. They are the human side of us, the part that makes each of us different from all the others, each equally alone, each equally unique, and each equally precious. It is by our feelings alone that we are made heavenly, by them we have known our hells, by our emotions we have been abandoned and resurrected and it will keep happening as long as we believe in ourselves, and in the meaning we have found in our lives. It is called the human condition, and we're all a living part of it. It may be a matter of neural nets, cross-associations, and molecular choreography, but it makes meaning and reality for us every day of our lives.

The painter Marc Chagall died at the age of ninety-seven. Like

Singer, he knew life's darker sides. Born in Vitebsk, a Russian Jewish ghetto that endured pogroms and persecutions, he survived the Russian Revolution and two World Wars. It was Singer's history, the very same, yet Chagall saw an entirely different world, one filled with childlike joy and faith in the human spirit. As he once told an interviewer, "There is no secret about it. You have to be simply honest and filled with love. When you have love all the other qualities come by themselves." It was the only world he knew, and he made it, and celebrated it, and painted it in all the colors of the rainbow. There was no plan needed. There was just lots of love.

Where did I come from? What is it all about? Where is it all going? The questions of human existence have their answers - over six billion variations on a theme. What is the purpose of the universe? It is whatever we believe it to be and for whatever reason feels the best to each of us. Why are we here? Is there plan or purpose? Of course there is. We're fulfilling our destiny every day. If we are unique, we're also, each of us, the best example of our kind, surrounded by others of equal originality.

After all, what have we been doing all our life already but demonstrating that? We have made things happen. We have made others happy. We have conjured love and endured tragedy; we have been touched by joy; and we have been gripped by terror. We really have. Every moment of every day, in this world we perceive and believe in, we are the ones who make it conscious; we are the ones to give it meaning. For better or for worse, until death do us part, we make it real.

8

Adventure and Avoidance

Rats, Routines, and Social Media

Pleasure seeking must be, if nothing else, the most normal behavior imaginable. Pain seeking is called masochism and is usually treated as a form of mental illness. Looking out for a good time is the daily concern of all other living creatures on earth and a daily pleasure for many clever humans.

There are limits to everything, however. Carved into the walls of the temple of the Oracle of Apollo in ancient Delphi were two fundamental instructions: "Know Thyself" and "Nothing in Excess". They form the original source code of the fine old Aristotelian mental system we've been using in the West ever since. These ancient axioms have proven themselves time and again in the face of challenge from both sybarites and ascetics. If we get to know ourselves to the extent that we know our limits and don't overdo them, our mental systems will not fault out. Consciousness will not crash as long as we stay in the middle of its natural path. Once we get those two basic rules up and operating, everything else is applications software.

The only glitch in this program is that it neglects to mention just how we are supposed to find our boundaries. Until we pass a limit, we can't really know that we've exceeded it, and who exactly is going to tell us who we are? Self-definition seems to be fraught with personal bias, so whom do we choose to name us? Something outside the limits of our own sphere would be the only source objective enough to rule on our own validity, something beyond those limits we're supposed to keep within.

Coming up with answers to questions like these is what kept the oracle in business, and her advice wasn't always precise. A large portion of her working hours, it seems, were spent sitting over a natural gas vent, babbling in brain-addled intoxication. This giddy glossolalia was utter Greek to everyone but the local priests, of course, who downloaded her random data into something that sounded like advice and charged a fortune for interpretation and interface services. In our times, this sort of scam would be considered very much in excess of "Nothing in Excess." The oracle would probably be busted for drugs and her handlers put away for a variety of morals charges and gaming violations.

And yet, when observing the more expressive devotional practices of the major world religions, we consistently find something very similar. Swaying evangelicals, swirling dervishes, media mullahs, and mega-church power preachers seem to share the same beat. The repeated "Hallelujah" or "Thank You, Jesus" brings the same inspiration to some as "Hare Krishna," "Allahu Akbar," or "Amida Butsu" brings to others. African shamans speak in secret tongues, voicing their devotion in the same manner as praying Pentecostals. The spirit of the oracle is born again, all over the world, as more and more seek a deeper communion. For some, it may serve as a release from the stresses and limitations of day to day society; for others it becomes a personal bonding ceremony between the self and the Divine. All agree that whatever the motivation or the method, the faithful do seem to enjoy their many forms of transcendental pleasure seeking.

As our world cultures mingle, the themes keep reappearing these days. Millions are "Born Again," others strive to reach "Bliss Consciousness" or take "A Course in Miracles." From "Satori" to "Samadhi,", from the "uncoiling of the kundalini" to the "indwelling of the Holy Spirit," looking out for a good time is one thing, but this worldwide revival in religious practice is remarkable. What does a swing to the spiritual bring that is so sought after by so many? Have we transcended all the gradations and degradations of sex, drugs, and rock and roll to a genuine search for the ultimate connection, or is this accelerating trend toward commitment and communion just a new height of selfishness for "me" generations satiated on worldly thrills? Considering it from this perspective, the growing worldwide interest in the spiritual may be less a trend toward an evolving human consciousness than various cultural varieties of avoidance behavior. It could be seen as a growing desire to turn ourselves into unquestioning believers and celebrate the spiritual in a world too complex and threatening for the likes of humankind, a idealistic search for simplicity in a stressful and conflicted time.

Transcendental bliss, isolated by itself, is hardly a highly evolved state of consciousness. Many have heard about unspeakable tortures suffered by lab rats, and a few have probably also heard about the unspeakable joys of some other lab rats some years ago. These latter rodents had electrodes implanted in the hypothalamus, a critical part of what is called the brain's "pleasure circuit." It was a pure connection to joy, and each had a button to receive as many peak experiences as desired. Oh, rapture! The rats would sit on their buttons until they dropped from exhaustion. They liked it better than sex, they liked it better than drugs, they even liked it better than eating. Normally, animals haven't the first vestige of organized religion, but this was the cult of the button for sure. They were converted on the spot, blitz consciousness, slain in the circuit, and fulfilled with all the ratty forms of joy. They surely saw the light, but the gods of their current redemption were much more interested in their hormones than the rodent hallelujolt chorus. Transcendental bliss, at such basic levels, is both predictable and easily reproducible.

Brightening our brain centers with that sort of impulsive ecstasy wouldn't be difficult. Human reaction is more variable, but with enough work, who knows what refinement could bring? It is just hard luck for the priests of the button that normal humans have never elected for this sort of brain surgery; and less invasive procedures, such as electroconvulsive therapy, do little better than overload the system and scatter the signals randomly. This can often derail a serious depression, but nobody has described ECT as a pleasant experience.

Recently, advancing technology has accelerated advances in a number of experimental applications making use of external electromagnetic therapies, others rely on more accurate placement of electrodes surgically. In 2007, Felipe Fregni, MD, of Beth Israel

Deaconess Medical Center in Boston, published the first randomized study involving electromagnetic therapy with forty patients suffering from recently untreated major depression. Forty percent saw a lessening of symptoms. More recently, Patricio Riva-Posse and Ki Sueng Choi, following pioneering work by Helen Mayberg at Emory University, developed a new imaging approach to more precisely pinpoint bundles of nerve wiring in the subcallosal cingulate region that seem to produce immediate relief and long-term antidepressant effects. There will doubtless be improvements, but until the FDA-approved radar good-wishing wand appears on the market, most people will continue to utilize the second part of electrochemistry, heading for chemo rather than electro. It can do the same thing and it's much more subtle. Whether it's a second glass of wine to put a glow on the evening, the touch of a special fingertip, the power chords of J.S. Bach or a concert crowd singing along with Springsteen, we have many ways to travel.

As adults, most of us have learned a variety of culturally acceptable ways to alter our brain chemistry and our state of mind at the same time. What we seem to be seeking, or at least what we seem to be getting from these activities, are levels of desensitization and personality generalization to allow for a greater communion with the other who or what. From the glad harmonies of music and the blurring effects of alcohol and social drugs to the very personal and persuasive rhythms of erotic sensuality, we are taken beyond ourselves to momentarily become a part of something larger rather than the solitary soul we know so well. Outside our own self-definition, with momentary out-of-our-mind perspective, we may gain those essential insights to judge our limits and help find ourselves in our own space and time. Those insensitive to the limits they discover may abuse themselves and others, but within bounds, even intense pleasure seeking is acceptable. Throughout history, our success is generally matched by the amount of pleasure that we take in any field of endeavor, from mothering, to managing, to meditating.

If we could wire up rats to create pleasure, perhaps we could wire up some monkeys to monitor what is happening in the brain when the feelings we associate with pleasure are taking place. In fact, they put the wires on monks instead, both Hindu swamis and Tibetan lamas. In the 1970s, psychologists Elmer and Alyce Green, of the Menninger Clinic in Topeka, Kansas, traveled to India to observe and record the unusual physical and mental powers of trained yogis. One such adept, Swami Rama, was invited back to the Greens' laboratory. There, under strict laboratory control, he repeatedly demonstrated extraordinary ability to influence and control both his body and his brain activity.

Ironically, the Menninger Clinic is located only a few miles from the very evangelical church where, at the end of a "Watch Night" service on January 1, 1900, the first American Pentecostals spoke in tongues. The Christian ecstatic tradition, a new form of emotional and spiritual release, began in Topeka and has since spread worldwide among many branches of the Christian faith. Swami Rama was close to home in many ways. More recently, the Dalai Lama invited researchers to study some experienced Tibetan Buddhist practitioners. Observing brain and body states during intense meditation, the researchers were able to verify that a number of monks were able to voluntarily control aspects of basic metabolism, from lowering blood pressure to increasing body heat, through purely mental exercises. Within a short time, versions of these practices were being made available in Boston, Massachusetts, as part of an innovative, hospital-run stress management program associated with Harvard Medical School.

As advancing technology brings us closer to actually observing the mind in action, we are beginning to gather some hard data on the states of mental grace. We have quite a way to go, but we are getting closer to a point in time when effective techniques may be commonly available to anyone who wishes to practice them. It would certainly go a long way toward establishing the sort of global human warming we so desperately need. If enough people could find an inner joy, it would be difficult to sustain any form of human-against-human suffering. Despite our proud posturing and bloody history, we are still primates: curious, clever, and skittish, but hardly bloodthirsty by nature. Given the chance, all of us just want to be happy and not suffer. If rats can do rapture with amps and Tibetans can do it with chants, it probably isn't that complex. Taking a careful look at the varieties of bliss, the first question to ask is, "What is happening in the brain when 'happy' is taking place?" What, simply stated, is actually going on up there when someone is in a state which they later identify as pleasure, joy, or ecstasy? The second question might be, "Is it enlightened or escapist to seek and enjoy such intense pleasure?" Finally, "Do mystical or religious experiences signify holy contact or just mental meltdown?"

Chemicals for Courage:
The Rapid Response System

The first question is the easiest to deal with in terms of general brain biochemistry, and it was previewed in some detail in Chapter 7. From a molecular viewpoint, we are doing it with adrenaline, ACTH, serotonin, dopamine, and a few other brain chemicals. The recipe differs from location to location, but nearly all physiological and perceptual effects of a state of bliss can be directly or indirectly associated with the effects of these hormones and neurotransmitters on some organ or brain structure.

Adrenalin, a major factor in all emotions, is a master hormone of the human body; it is our natural metabolic catalyst that can in turn affect the release of numerous other compounds which can have dramatic effects on the brain and the body. It is produced in the adrenal gland, a clump of chromaffin cells located at the top of the kidney. As "kidney" in Latin is *renes*, ad-renal-in simply describes where it's made, "substance from the kidneys," a handy place to get it into the bloodstream. The adrenal gland is a little local chemical plant with wires up to the brain. Like a personal overdrive, occasional adrenaline is normal, but too much can exhaust or impair us. Many are aware that highly adrenergic states, if repeated enough, lead to stress, anxiety, high blood pressure, drug and alcohol abuse, and impairment of the immune response.

On the other hand, as adrenalin is a better stimulant than any drug yet devised, it might have been expected that we would work out ways to get a little more here and there. Bears don't mind a few bee stings if they can get to the honey, and humans have always been looking for various ways to unwrap themselves from their overly personalized, solitary minds even if some of the techniques are not only daring, but dangerous. It seems that, over time, we learned how to manipulate our endocrine system to produce a variety of consciousness-altering experiences at will. It has to do with how internal brain activity can initiate hormonal overloads and how stone-age shamans hacked the code ages ago, probably by mistake. We unconsciously learned how to create experiences guaranteed to expand our consciousness and, at the same time, anchor us more realistically in the present.

An easy way to unload the adrenal gland is to register a sudden increase in brain activity. This would indicate that something either very familiar or very dangerous was occurring out there. As a result, the trip signal is a sudden rapid rise in the uptake of the neurotransmitter norepinephrine. Epinephrine is simply Greek for adrenalin; in Greek *epi* means over and *nephron* means kidney. However, since "epinephrine" sounds more scholarly than "adrenalin," noradrenalin is now called norepinephrine, the "substance preceding adrenalin".

Norepinephrine is found in much of the neocortex, where most of our recently evolved brain mass is located. When norepinephrine is suddenly being released all over the place, it's like yelling "fire!" down the wires, and the adrenal gland obligingly dumps the adrenalin into our bloodstream without asking what's happening. Usually, there isn't time to ask, which is why it has to be so automatic. The speed with which this electrochemical circuit can launch is remarkable. It is, after all, our emergency getaway system, designed for instant deployment in life-threatening scenarios.

On any given day, for instance, we are walking down the street. Our body is operating well within its limits, our senses picking up information and processing it at normal brain speed as we maneuver through our regular day-to-day consciousness, virtual interfaces smoothly guiding us through time and space. Then suddenly it happens. It doesn't matter what happens, it just has to be dramatic enough to elicit a sudden release of norepinephrine. In this instance, our pleasant stroll is shattered by the squeal of brakes shrieking into our auditory cortex together with an exploding image on our retinas of an out of control car skidding into the intersection just ahead of us.

As information from our eyes and ears hits the top level of the brain, neurons race to compute escape trajectories. There is a mild biochemical eruption at the higher cortical levels as neurons careen into overload for a microsecond. The sudden flood of norepinephrine into synapses interrupts the lazy spontaneous firing of resting neurons, opening the gates for more sensory information pulsing in from outside. Neural circuits flash-energize like neon signs, extending in a crackling frenzy faster than thought, as patterns and networks start interconnecting like a fireworks finale. The chain reaction shifts the entire brain into electrochemical excess. In less than a tenth of a second, the surge hits the adrenal gland, completely bypassing the brain's interpretive centers. In a flash, the adrenalin unloads directly into the bloodstream, rushing up the arteries to a brain that is only a heartbeat away. Like gasoline dumped on a blaze, it hits the frantically firing neurons, energizing both hemispheres for a massive reaction.

Now the visual cortex at the back of the brain swings into

power overdrive, neurons firing away at triple speed like that movie camera suddenly cranked up to the max. Both filmmakers and brains can create slow motion visuals, that almost miraculous sense of "time standing still" when we really need a miracle to get us out of harm's way. With the brain on fast, and our world in slow motion, all the smooth muscle in the body is contracting at once. Hairs snap straight up in their follicles, the diaphragm contracts in a gasp, and all the blood vessels give a squeeze, sending a surge of blood toward our liver and a shiver through our body. We are now in a state about as close to direct contact with the world around us as we will ever be, but there's no time to reflect on it now. Less than half a second has passed; already the liver is jettisoning its rich supply of glucose into the blood, charging it and fueling the muscles for powerful and immediate action. The body sweeps through biochemical transformations faster than the mind can think in sequence, slamming molecules into position, wrenching the spine straight, pulling time itself to bits as life and death hold fast for split second while we yank ourselves out of danger in the nick of time.

In the full adrenalin shock of trauma or terror, mothers have lifted cars off children, hunters have leapt to impossibly high branches, and accident victims have walked with broken legs. Adrenalin is our shock-action hormone; if it kept on coming, we would drop from burnout and exhaustion. Under circumstances such as these, it's hard to recognize a pleasure circuit when we see it. Getting our pants scared off is not anybody's idea of ecstasy, especially when it's real danger. Adrenalin makes everything *right now!* Excitement perhaps, but hardly rapture in any recognizable form. If the car skids out of the intersection, or if it turns out to be just some kid laying rubber for kicks, the hormones taper off immediately. The smooth muscle in the body now relaxes, with predictable results. The cold shiver is replaced with a warm rush as blood returns to the extremities. At this point, the bladder or bowel may fail if their sphincters relax too much; a not uncommon occurrence during a terrifying experience. There is a warm hormonal tingle throughout the body as the brain re-starts cognition in a foggy biochemical afterglow.

It will take some time to cool down, synchronize the parts, and restart the prefrontal sequencer to regain our measured, rational, comparative, and predictive thought again. We were yanked beyond our limits, and lived to talk about it. There are always artifacts as well: brilliantly remembered moments when we were so much a part of the world around us that we were nearly out of our minds. If we had been lost in our personal virtual chronology up to that moment, it was a sudden drop-kick into present tense perception with the volume up to twelve. A full-strength adrenalin rush will always slip a new card into the deck. The future will never be quite the same.

We may go through this a lot more often than other primates because of what happened during the last two million years as our brains evolved and enlarged. In evolutionary time, it can take ages for major adaptive changes, for a nose to evolve to a trunk for instance. Evolution may occur by fits and starts but it does not happen quickly; manipulation of genes is involved, and humans take at least fourteen years to recycle a set. The concept of redesigning a whole nervous system, our entire hormonal checks-and-balances operation, with a million year deadline is not feasible unless we're all creationists; interrelated systems change together. It would be hard to imagine talking, for instance, if our tongues expanded their mass fourfold without some extensive dental work and a complete jaw retrofit. "Add more brain tissue" was a simple update, and one gene did most of the work. We managed to quadruple the volume of the brain in less than two million years. Chronological recall, abstraction, and projection were overnight in comparison, probably less than seventy-five thousand years ago.

Either way, our hormonal systems originally operated with the standard 300 cubic centimeter brain that guides the average primate. Within those two million years, however, we managed to upgrade the old 300 cc standard issue to our mega-memory, parallel-processing, pattern-sequencing, 1,400 cc electrochemical powerhouse. It was like bolting a Harley Davidson motorcycle engine to a skateboard. The endocrine system had been accustomed for millions of years to taking orders from less than a quarter the number of neurons we now had howling "wolf" or "wonderful" when anything popped up on the screen bigger than bite size. We can get excited by nearly anything if we set up our hair-trigger human consciousness with enough repeated stimulation. This is not to suggest that packing on the last 1,000 cubic centimeters of virtual memory and processing capacity was a bad thing. With the added cc's, we gained time, space, imagination, and reflective conscious perception, but our brain still operates by electrochemistry. From an informational point of view, our greater neural mass is just more processing power and virtual memory. From an electrochemical point of view, however, four times the amount of biochemical turnover can be created for the same amount of stimulation. As a result, our endocrine system is getting wagged by the brain.

And it wasn't just the higher cortical centers that grew. The fine-movement-controlling cerebellum gained proportionately just as much mass. This was important, because fine-tuned sequential muscle response had originated to let apes and monkeys navigate about in the treetops. Once again, humans ended up with nearly four times more. The medial-forebrain bundle grew to transfer subtle prefrontal motivational data directly to the brainstem itself; humans can intend to do it right the next time. Our moves are much more mindful. Many brutes have more brute strength than we do, but humans are the smoothest on earth at any learned form of muscular coordination. No monkey can match the aerial grace of human gymnasts; cats couldn't dance if they wanted to.

As we woke to chronology, the recollection of the seasons allowed us to predict, to prepare, and to do the one thing that nature never does and never can do. We learned to repeat on purpose. It was the beginning of our control over our environment, our control over our destiny, and our final evolution as a species. We learned to plant the fields, to net the fish, to fire a clay pot. Weaving, one of our first technologies, was developed by women, and portable looms were the first technology transfer. We learned how to try, predict, and try again. The world will never do it again the same way, but we learned to repeat first for mastery, and later just for fun.

Our cognition became proactive as we began to invest meaning into whatever we did the most, and we started to re-shape the world into a mirror of our likes and dislikes. Our neural nets grow even as we sleep, familiar networks extending as random pulses wander through them while we dream. This constant activity and change is what makes it impossible to compare the brain and any sort of computer. Like a quantum flash of time that theoretically locates an electron in a cloud of probabilities, the images of brain patterns changing and shifting through the physical channels and chemical bridges may be discernible only to a hidden impulse in an otherwise chaotic mass of streaming activity. We can only imagine that which our own mind can focus, and these possibilities are beyond the focusing resolution of the only consciousness we have.

In 2016, Northeastern University physicist Paul Champion and his co-authors found that protons in cells are able to quantum "tunnel" at room temperatures, acting as waves inside the confines of a living organism, supplying their energy to mitochondria. Other hard-to-observe quantum effects may be hard-wired into basic biological functions, happening so fast, or hidden in a cascade of other reactions, that they fly under the radar of most experiments. We learn more each year, but there is far to go. As we age, through repetition of familiar activities, our incidental patterns eventually so extend and interconnect that nearly any familiar stimulation can awaken a large associative network of neurons. Many neurons are often on the edge of response simply from the stimulation of internal chatter and are easily pushed into heightened activity by a dose of external perception.

Facial recognition is quicker in women than in men, but either way, the visual data takes the medial forebrain bundle shortcut directly to the hippocampus, a part of the limbic system, the basic structure linking memories and emotions. Since we judge reality by emotional clues, the limbic system serves as the immune response of our personal consciousness, our personal virtual reality check. Due to the large store of memories, facial recognition alone can produce quite a surge of emotional reaction even before we start up the cognitive search-and-compare processes. The hippocampus got its evolutionary start directly from the olfactory bulb, which accounts for how our emotions are easily aroused by familiar scents. Smell recognition completely bypasses the brain's fine-tuned arousal filter, the reticular activating system. It often surprises us when an unexpected scent evokes a vivid rush of emotional memories. The activity of the brain going about its business may be an energetic enterprise, but its basic structures cannot be subjected to constant excitement. If the hippocampus is overstimulated for too long, its cellular mechanisms will start to fault out from exhaustion. If things get too excessive, in other words, it loses control, and the system can shut down. If this happens, our virtual reality, our sense of the world and its meaning, will destabilize.

As the limbic system defines reality for each of us by denying emotional existence to all but parts of our moment-to-moment perception, it also provides the boundaries between thought and belief. It is limbic system distortions, along with suspension of chronological time, that create many of the characteristically altered perceptions common to dreams, delusions, and near-death experiences. Although our sense of reality may start in a part of the brain as old as the hills, it's not hard to hot-wire the system if the situation is right. The majority of our memories are unconscious most of the time, but we must all be subject to specific and personal hormonal responses to innumerable old, embedded and forgotten patterns. If they are memories we treasure, recollection and thought will intensify and further personalize our responses. Our ability to amplify experience with memory, often triggering pleasurable hormonal rewards, is a major factor in how we get our bearings in life and make the world familiar to us. Of course, there are drawbacks. The massive mind we depend on for everything operates within its own virtual reality. If we don't periodically re-evaluate our repetitious accumulations, we can begin to believe our internal world is the only one there is.

Over time, personal habits make us ever more repetitious and insensitive to the changing conditions around us. This is one reason that both the Internet and the smart phone, far from expanding the reach of the user's mind, can allow us to repetitiously recirculate our preferences through texting, video streaming, and following selected social media. George Saunders, in a 2016 *New Yorker* article, drew a focus on this tendency. "In the old days, a liberal and a conservative got their data from one of three nightly news programs, a local paper, and a handful of national magazines and were thus starting with the same basic facts (even if those facts were questionable, limited, or erroneous). Now, each of us creates a custom informational universe, wittingly (we choose to go to those sources that uphold our existing beliefs and thus flatter us) or unwittingly (our app algorithms do the driving for us). The data we get this way, pre-imprinted with spin and mythos, are intensely one dimensional." An entire generation now runs the risk of turning into virtual selfies of themselves, involved in a virtual world within a virtual world.

How, then, could we get a better look at the present moment to experience it as it really is rather than distorted by layers of recollection and expectation? How can we get a really new perspective? Why, simply let loose those hormones and shoot into psychosensual overdrive for a moment. If we energize enough of our associative memory, we can amplify any experience. Then, with limited amounts of external involvement, we can have all manner of meaningful excitement right between our own ears.

Associative avalanching can trigger major hormone releases in people for reasons ranging from budding romance to serious scholarship, from a celebrity on the stage to a cockroach in the kitchen. The human mind has harnessed the world to our whims, but individually it can propel us into some of the most inane sorts of behaviors. With age, we only become more specific. Teenagers swoon en masse at the same pop stars while adults, having grown more personal, have more specific heroes in their own larger, but specialized, fields of association. Cultural holidays and family events can be guaranteed to evoke strong feelings in us no matter what our race or nationality. Our most basic memories, the earliest ones, are those unite us all, and we all feel them the same way.

Enmeshed and entrapped in a world we have made, we seek this refreshing stimulation because it is that necessary and repeated meeting with our hormonal cues, forcing our feelings to their natural limits in each of us, that alone can keep us defined and alert in a world we have made much too safe for such a curious little primate. Sometimes the trip circuit is unconsciously planned into our work routine, such as never allowing enough time and getting caught in exciting panic deadlines much too often. Some hobbies, sports, and occupations are clearly going to include more thrills than others; police officers, race drivers, and rock stars wouldn't do it if it didn't excite them. As rock-climbing outfitter Yvon Chouinard remarked, "Doing risk sports taught me another lesson: Never exceed your limits. You push the envelope, and you live for those moments when you're right on the edge, but you don't go over. You have to be true to yourself." Still, nobody ever lost money with a good roller coaster. Fortunately, most of our personal routines and repetitions are not extreme, and we reach our hormonal happies less dramatically through varying forms of exercise, entertainment, or familiar ritual, be they domestic, cultural, or religious.

There is a darker side to all of this, however. The more we repeat anything at all, the more familiarity it will have, and the more we will tend to further repetition. Survival is why we have a brain to begin with; anything we survived is better than the unknown. The result is that anything from personal habits to destructive relationships can, and will, through repetition eventually become self-regenerating patterns. Soon they are altering our course through life like the invisible attractions of large planets, massive with our accumulated past, able to pull us again and again into old familiar orbits without our conscious will or even realization. By the time we've lived a number of years, we begin to repeat by unconscious habit those activities which were once quite coincidental. Our virtual walls are being erected without a lot of noise, leaving us emotionally with a rather limited space in which to exercise the full spectrum of our powerful systems of perception, thought, and action. Rarely, if ever, facing real life-threatening or life-enhancing situations, we become first acclimated and then dependent on reliable emotional triggers like a rat twiddling a button for safe little jolts to help forget some primal urge to get out and raid a grain sack in the real world outside.

Instead of being attentive to the world surrounding us, and its full range of possibilities, we are heading instead into the world of synthetic wins and imaginary rewards, rich with predictable pleasures. At the same time, even our most painfully unpleasant attention-getters are eventually woven in as part of the repeating and self-fulfilling scenario. "Honest" expressions of anger become repetitious conflicts against the same old adversaries; infatuation or excitement can be honed by compulsive attention to any person or activity. When human inventiveness in prediction and planning begins to be used mainly to arrange personal repertories of the same loves, the same hates, the same fights, and the same triumphs, it's hard to explain it. As a result, we will often go to great lengths to create and articulate personal, cultural, and even national myths to justify repertories of the same old patterns of expectation, pleasure, disappointment, and rage—the human hormonal four-step that too often replaces the harmony of life.

High Technology and Gridlocked Minds

Helping us along in this trend to replace the uneven experience of life with predictable thrills is the enormous amount of sophisticated technology currently deployed to create and promote events and entertainment intended to excite us. Today, the promotion of augmented realities, virtual realities, professional sports, fantasy sports, video games, "reality" shows, movie stars, Star Wars, drug wars, news bites, super heroes, and sitcoms is a multi-multi-billion dollar industry.

As soon as our modern civilization saved us from being faced with real danger, it seems we had to spice our lives with ritual romance, drama, and mock involvements to save us from terminal boredom. As the Roman poet Juvenal lamented in 100 AD, "everything now restrains itself and anxiously hopes for just two things: bread and circuses" - processed food and cheap thrills. There is nothing new here; we've simply brought the Coliseum into our lives so that mayhem for the masses is more profitable. Between vicious politics and religious wars, even civil agendas and global concerns become exciting spectator sports. At the same time, it's clear that an ever growing number of us feel disconnected from our technological twenty-first century world, a situation that simply increases our unconscious itch for some sort of deeper experience. On the good side, nations unite to feed the hungry, resettle refugees, and fight Ebola and Zika. But for the average individual, with the pressure of survival being settled, the search for something meaningful can mean the choice between a cable debate or streaming Netflix for a few hours of escape into another universe.

By 2018, advances in broadcast, cable, and Internet communications had already reduced much of the developed world to the level of a popular media-cracy. Policy is formed by politicians elected in response to the way they are perceived in the media as responding to imagined or exaggerated dangers. As wars and other perceived threats are the easiest way to get the hormones running, winning elections requires wars to be declared every year or so on foes as unlikely as refugees or opioids, renewed yearly as "The War on Terror" or "The War on Drugs" as well as war on every domestic ill from cancer to obesity. Dealing with everyday life challenges just doesn't seem exciting enough in a world filled with individuals who don't know why they're so anxious and don't know whom to blame. Meanwhile, technology-related distractions have been stealing our attention for decades: television instead of dinner at the table, video games instead of board games. There's an increasing tendency is to treat life like a variety show, quickly abandoning anything unexciting or boring, channel surfing our virtual reality for a quick fix. The broadcast networks once had only four money makers, "the four C's": Cops, Crime, Courts, and Comedy. The twenty-first century added popular entertainment contests, reality shows, and adult cable sagas lasting years with massive audiences. Even the news is biased by the search for whatever plays best to our most basic emotions, slanted and edited with specific attention to the sensational.

In one famous NBC memo, it was clarified that "news" should never be shown as it is, but packaged as little mini-dramas. We are informed mainly by practiced newscasters whose packaged candor and sincerity has reached new heights of hypocrisy and manipulation as they share with us the news that boosts their own network ratings. Over twenty five years ago, in 1993, veteran newscaster Dan Rather said aloud what many had been thinking. The problem? "Our bosses. They aren't venal, they're afraid of ratings slippages. They've got us putting more and more fuzz and wuzz on their air, cop show stuff, so as to compete not with other news programs but with entertainment programs — including those posing as news programs — for bodies, mayhem, and lurid tales. We trivialize important subjects. We put video through a Cuisinart to come up with high speed cross cuts. And just to cover our asses, we give the best slots to gossip and prurience. We should all be ashamed of what we have done." Caught between politicians driven by the need for press and media conglomerates who need more viewers, most of us are getting a slanted, edited, and often destructive view of reality from all sides. Nothing sells like warnings and promises, and so it continues day in and day out; pleasure jolts from the media priests of processed, amplified, and endlessly promoted versions of the world around us.

Those who suffer the most, as in any wars, are the young. Allowed to park their minds in front of screens for longer and longer periods of time, their brains are rounded by ratings and their reality can becomes permanently influenced. In June, 2016, the American Academy of Pediatrics issued a recommendation that children under the age of six be shielded from on-screen violence. "Toxic stress" caused by racism and violence can take a heavy toll of the learning, behavior, and health of children," wrote Dr. Jack Shonkoff, Director of the Center on the Developing Child at Harvard University. Such stress, if continued, can affect brain development, the cardiovascular system, the immune system, and mental health. "We are jeopardizing children's lifelong health," he added. A few month later, researchers from Brown University determined that children who spent more time watching TV, playing video games, or using a smart phone were less likely to finish their homework and, perhaps more ominously, showed less interest in learning generally.

Reacting to the study, Dr. Ellen Braaten, Associate Director of the Clay Center for Young Healthy Minds at Massachusetts General Hospital, was not surprised. "It makes sense. We've said for a long time the more time kids spend on electronic devices, it's probably not that great for them, but we didn't have a lot of data. This study validates that idea that other things aren't getting done." By now, millions are being molded not by history, but by edited media versions of it. Past events are replaced by fictionalized dramatizations. "We live in a media age," remarked film critic

Leonard Malkin. "If a television or theatrical movie can paint a vivid enough picture for young people, they'll believe that's the way it was."

We are surrounded by so much virtual hype that it has begun to inhibit our ability to enjoy the simple congeniality of a so-called normal existence, paradise for any lower creatures. This situation can be blamed almost entirely on the private media, reaping fortunes while force feeding our senses a deadening diet of danger, drama, diversion, and pharmaceutical advertising. As a result, the real world dulls. We begin to use it mainly as a staging ground for dreams and fantasies, constructions we weave from synthetic hopes and popular fears and carry about in our minds helping us to create a world as two-dimensional as any talk show or infomercial as we become the stars of our very own viral video. Eventually, self-generated cycles of predictable thrills and chills can build such barriers that even instances of real human drama, tragedy, or joy lose their power to teach or guide. It would be naive to expect that one or two full-tilt hormonal rushes could unknot a mental harness perfected by years of habit. In most instances, those seeking peak experiences for insight and discovery are too often only temporarily illuminated, and the glow usually dies away. Others rapidly converted to the newest trends are often as rapidly unconverted as deeper and more personal tides of personal repetition wash away at their determination to remain open and responsive. Too often we fall back into comfortable and familiar patterns of our virtual joys and synthetic fears, punctuating and perpetuating our everlasting oscillations in a world of echoes and dreams. Our neural nets, there to save us from need, instead grow untended like kudzu across our internal landscapes until we are nearly strangled in tangles of vain illusion and imaginary fears. These days we have developed ever more novel ways to shield our senses from the present moment. Many become completely satisfied with near total personal avoidance of the uneven experience of life, replacing it with routines as easy to maintain as the rat's pleasure button.

Ironically, new and unexpected technologies may have actually accelerated this trend. With over a billion smart phones in circulation, their effects on society are beginning to feel as ominous as global warming on the weather. According to a 2015 Pew Research Center Poll, our pocket portal to a personal patchwork of Internet worlds is starting to affect basic interpersonal communications. In the United States, with 79 percent of the 30-49 and 82 percent of the 18-29 age groups using FaceBook and other social media sites, large numbers of people are avoiding much of the world around them nearly entirely. Studies have shown that even the presence of a smart phone - just lying innocently on a table – can affect the emotional and empathic depths of conversation. As Marianne Curcio wrote in the Boston Globe, "At some point in recent history it became perfectly acceptable to have a face-to-face conversation with people while they are looking at their phones. They can hear you because they are murmuring responses and nodding their heads in the right places – but their soul is given over to the screen. You wouldn't read a novel or paint a watercolor while engaging in conversations with someone, but it's somehow OK to check your social media feeds."

Encircled by virtual friends and personal social networks, some become so dependent on the digital fix in their pockets, their limitless gateway to a world of possibilities, that they cannot focus on the real world unfolding before them. In a recent letter to the editor, a Boston writer described a scenario that summed up the current situation. "At a park not long ago, I saw a young girl give a flower to her mother, who took it without glancing up from her phone. Nearby parents and their two children sat in silence, each playing Pokemon Go." Silicon Valley startup guru Tristan Harris went one step further, confronting smart phone syndrome as a troubling new addiction and promoting strategies to avoid the entrainment it fosters. In a 2016 *Harper's Magazine* article, he observed, "Our generation relies on our phones for our moment-to-moment choices about who we're hanging out with, what we should be thinking about, who we owe a response to, and what's important in our lives. And if that's the thing that you'll outsource your thoughts to," he added, "forget the brain implant. That is the brain implant. You refer to it all the time."

Subtly at first, and with greater regularity as we become habit-ridden or Internet entrained, our emotions can become stunted and ritualized. We can become as self-engrossed in our self-created sideshow as any gaming addict, hunched over his controller, flashing and blasting in his own dark arcade. Like rapturous rodents with a wired-up brain, we turn increasingly to well rehearsed, reliable forms of self-stimulation, eyes turned inward to our hopeful fantasies, waiting for the jackpot we have learned to fashion from our own mental substance until we become trapped in a virtual reality with virtually no way out. As our natural tendency to follow past experience through future expectations into various mind-deadening states of mental gridlock seems to be unavoidable with an accumulative memory, it is no wonder that we seek an escape.

Our temporary solutions are to shock the mind into states of thoughtlessness with the help of various social substances, excessive sensuality, and thrilling ritual. These will always be available. If there were any permanent solutions, however, or at least some practical methods for a thoughtful dip into the present tense without self-denying or self destructive behavior, it would seem whoever has them should step up and tell us about them. Actually, people have been doing that for ages. We call them prophets, saints and saviors, and we call their suggestions "wisdom."

Can we cleanse the brain of its accumulation of outdated pasts and one-sided futures without getting brainwashed in the process? Can we straighten out the tangles in the neural nets we created ourselves, those cycles that keep sending us in circles, without losing our bearings or our brains? In fact, it is not only possible, but numerous ways to achieve this end have been taught and practiced throughout history as spiritual instruction, mental training, personal skills, and even more powerfully, embodied within the most fundamental rituals of traditional religious practice wherever it is found.

Part Four:
The Future

9
Priests and Prophets
Fulfillment in Real Time

At any moment, our perception is anchored in the stabilizing neural illusions of memory and expectation. Our cognitive reflection operates in chronological time because we think comparatively. We compare ourselves against outside events as well as inner projections to form our self-image. This requires the ability to continually scan patterns, shuffle sequences, and compare them. If we lost that chronological underpinning, our situation could become incomparable; placing us in pure emotive perception with no space/time limitations

In this virtual world of time and space, we need one in order to perceive the other. As a result, whenever our sense of time checks out, the temporary self goes with it, and we get a chance to experience life from a truly universal, unlimited perspective. In timeless perception, we always forget ourselves. This is no state of mind for doing the taxes or even crossing the street, but it is superlative for a sudden shift into the immediate moment, an excitingly direct interface with the real world. It is considered a form of higher consciousness in the East, where getting totally outside the personal context is viewed as the major goal of spiritual endeavor. In the West, the experience of a state holy grace is equally beyond personality or particulars. So how do we do this?

Fortunately, we don't have to unplug the brain and pull out memory chips. Chaotic quantum pattern memories aren't built like that, and the techniques we use are quite different. Just as time distortions occur when the brain speeds up, there are ways to create even deeper experiences if we selectively distort normal consciousness through self-induced neurological routines. In this manner, rehearsed and arranged electrochemical anomalies can lead to states of consciousness ranging from mild hormonal highs to delight, amazement, and even ecstasy.

Planned Satisfaction: Stretching the Nets

The interconnected nature of the brain seems to guarantee that associative networks are interlinked throughout many different structures, looping through discrete areas that manage different tasks. One way to unhinge memory, then, could be to overstimulate certain parts of the system and unbalance normal brain activity. We can selectively knock out parts of cognitive process if we exhaust our neurons by overloading them until they simply can't react properly. This is similar to what Internet providers call a "denial of service attack," the log jam that occurs when hackers gag a server with too many requests. If we could trick our brain into doing this, large networked patterns stretching throughout the brain could be unraveled temporarily as the needs of one part of the system shift out of synchrony with others. The detail of the holographic quantum patterns could easily drop by a factor or so. Ego could slide out of focus without losing attention.

For instance, if we over stimulate and disrupt functions in the higher cortical areas, we could temporarily "crash" our perception into a lower-brain backup, a form of low-definition consciousness. The cerebellum might keep on clicking away, but the neocortex would be temporarily disoriented. This could briefly drop us back into something much closer to instinctive sensibility for a moment. We would be unthinking and yet aware. These highly charged forms of selfless perception, a state of grace to a religious Westerner, *samadhi* or *satori* to a Hindu or Buddhist, is actually the same place, a self-induced state of higher consciousness, lower consciousness in fact, and we all have the ability to do it. In fact, we've all done it many times.

The usual method is to create large synthetic patterns in the brain through attention to planned activities that are largely repetitious. This virtually insures that the memory and the operational controls in the cerebellum are becoming linked. There will eventually be enough associative neural networking to trigger a hormonal response. Since adrenalin release will speed up parts of the brain while intense repetition will exhaust other parts, if we keep both of them up long enough, comparative perception will occasionally short out as input-output faults begin to appear at various levels of the cortical structure.

Eventually, mindful repetition of any complex activity can trigger enough response for mild hormonal body highs, producing pleasant mental and emotional stimulation without either fantasy or frustration. It's not a virtual event, it's a real world experience, and the added hormonal surge greatly heightens our experiential perception. Creating such large networks takes time, however. Unless we are into Eastern meditative practices or artistic obsession, we are not by nature very good at perfect repetition. In fact the way most of us achieve these states takes advantage of combining differing brain areas simultaneously for an effective overkill. We can do this with almost any mind/body activity which requires thoughtful practice, from music to yoga, from drama to dance.

In a 2008 article in *Scientific American*, a Columbia University neuroscientist posited that synchronizing music and movement constitutes a "pleasure double play." Music stimulates the brain's reward centers, while dance activates its sensory and motor circuits. Regions of the brain that contribute to dance learning and performance include the motor cortex, somatosensory cortex, basal ganglia, and cerebellum. The motor cortex is involved in control and execution of voluntary movement. The somatosensory cortex in the mid region of the brain generates the sense of touch and plays a role in eye-hand coordination. The basal ganglia, deep in the brain, helps to smoothly coordinate movement, while the cerebellum, as expected, integrates and times input from the brain and spinal cord, fine-tuning complex actions and thoughts.

The ability of our forebrain and cerebellum to pre-plan these complex patterns of muscular movement evolved, in humans, into our unique ability to derive abstractions and predictions through sequential comparison of the patterns themselves. In the ongoing process of living, we will cross-connect all sorts of neural networks if we repeat any careful activity over and over. The constant repetition involved in the mastery of any skill occurs until we achieve a desired level of competence, which is an image we project in our own mind. As each repetition occurs in a different time frame, a new pattern is perceived by our senses at that moment. Since mastery requires practice, this will inevitably lead to emotionally charged experiences Unlike our mental gridlock patterns, unconscious habits we repeat and then try to justify, these are patterns that we created with conscious attention and full knowledge of what we were doing. Far from stressful, this time the feelings are usually marked with enthusiasm; a word again from the Greek – *en theos* – a feeling that God is with us and our endeavor.

This is how artists and athletes, every craftsperson and every musician, and anyone else who has experienced the personal glow of a job well done, creates those moments. When our skill and the circumstance combine in just the right way, we lose ourselves into the moment. Such moments of mild ego loss are instructive, not destructive, because they were sought purposefully. The late geneticist Barbara McClintock won her Nobel prize for discovering high-protein corn hybrids for a hungry world. For her, daily work with chromosomes was a social experience. "When I was really working with them, I wasn't outside, I was down there, I was part of the system ... these were my friends...they become part of you. And you forget yourself. The main thing about it is you forget yourself." What makes the experience particularly nice is that it often happens when we really are doing our best and are in the presence of friends or even admirers.

There are hours of dues to pay, of course, as we set up those deep patterns during days of mindful repetition. There are long hours practicing scales that miss, a dance that doesn't, the slapshots that slip, and the dunks that don't are all part of this patient assembly of those mental patterns that will let us lose our fears without losing ourselves. Over time, amateurism becomes expertise. The body begins to move in smooth curves of carefully controlled energy. The fingers find the frets without hesitation, the colors hold, the dancer's body wakes, and the energy begins to flow from within.

Sooner or later, the experiences start to happen. Practice and performance are finally in tune; the puck slides into the net, the ball soars over the goalpost, and the moments all musicians know when the music takes over, sweeping them into harmonies as mind and body momentarily forsake time and space in the glow we know so well. In *Flow: The Psychology of Optimal Experience,* psychologist Mihaly Csikszentmihalyi describes the moment when one is "completely involved in an activity for its own sake. The ego falls away. Time flies. Every action, movement and thought follows inevitably from the previous one, like playing jazz. Your whole being is involved, and you're using your skills to the utmost." This is full integration in the flow of life, the Tao, the Dharma, God's kingdom, and happy to simply be here. With our attention redirected from our internal cycles, we rediscover our spirit alive and well in the world we knew before self and ego. Whenever our mind becomes fully involved in the practices we enjoy and repeat, our attention shifts from ourselves to a full partnership with the world around us, only a tenth of a second away, and the distance seems to vanish.

For the 350 million people around the world who suffer from depression, this may be a time of new insights. A long known but under utilized form of therapy, Behavioral Activation (BA), which is based on principles that echo Schmahmann's observations, was recently shown to be just as effective as the most popular approach, Cognitive Behavioral Therapy (CBT). CBT, which benefited from rigorous scientific studies demonstrating its effectiveness, emphasizes learning new thought patterns. BA combines this with activity. "The idea is that what you do and how you feel is linked," says David Richard, a health services research at the University of Exeter in England. "If the patient values nature and family, for example, he might be encouraged to take a daily walk in the park with his grandchildren. This can increase the rewards of engaging more with the outside world, often a struggle for the depressed." One advantage, noted in his online study in a recent issue of *The Lancet*, is that it is easier to train therapists in BA than CBT. Doing what comes naturally, and enjoying it, can actually put us back on the tracks.

To a certain extent, then, every time that we thoughtfully repeat something that we love to do, we add to our growing networks of associative energy. Then, when outside events energize them within the right setting, we may find ourselves experiencing a very pleasant hormonal whole-body glow, accompanied by partial ego loss in supportive and protective surroundings. Of course, the more brain mass that can be called into resonance, the stronger the feeling. It takes a lot longer if we just watch. A full involvement in life and the things we love to do makes it so much easier. Every time we lose ourselves in our music, art, hobbies, studies, athletic contests, professions, personal fitness, volunteer activities - even cooking, working, and parenting, we are merging life with our virtual reality and making us a part of a bigger pattern. Since the joys of personal fulfillment always require practice, it's important to find a practice that we can enjoy. If paying the dues is a pleasure in and of itself, the payoff will come sooner and will be even more pleasurable.

Harnessing the Tools of Transformation

There are, moreover, even higher power versions of these methods of finding an eternal moment. Once again, the basic technique is the same, but with added subroutines that tend to act as catalytic boosters, extra mind exercises that kick in at just the right time, adding such direction and power to the experience that the effects can last for days, months, or even longer. In Christian, Muslim or Jewish settings, these experiences are associated with religious rapture. The Hindu might define it as shaktipat, and an African shaman would suggest holy possession. When these experiences occur within the focus of a regular symbolic or physical practice, they can provide a feeling of fulfillment both profound and indescribable.

In fact, the only other time that we could really be so purely fulfilled was in early childhood. As infants, we were the center of the universe; everything was done for us and because of us. The world loved us completely. We will never again experience the sure and undifferentiated self that we knew as very young children. Comparative cognitive consciousness won't even operate until the systems architecture is solid; we spent a long time in a place where time didn't count for much and we were the only soul that mattered. Like Adam and Eve, we assumed that Eden was here just for us, and would always be our home.

This universal time of early innocence is, in that sense, also a time of profound self-knowledge. It is our human misfortune that this particular form of perception is nearly impossible to recapture or reproduce in the mature human brain. Even if we were to recapture it, we could only use it for the experience itself. Noncomparative, asynchronous thought may be the experience of being in a sureness we have sought ever since, but it is what it is, an immature and unorganized form of our precise and measured consciousness, a baby's view from an infant brain. Still, it was the last time our mind was at one with our world, and we need the reminders.

To shake hands with our soul, then - that part in each of us that does not think, but knows who and what we are - all we have to do is to recreate that infantile brain state and have a reunion with ourselves beyond comparison, before good and bad, before right and wrong. In those endless days we just "were," and we didn't worry much about it. It had been that way forever. If we want to find ourselves as adults, this is the way to go, but it is not a transcendent move upward. It is a return to another reality, the one that we knew before we knew anything else, when our world was there only to comfort us and love us.

To achieve these states we must do more than create some short circuits with hormones and planned neural saturation. Only through deep and massive disruption of the circuitry of the brain will multiple parts of higher consciousness fail at the same time, creating momentarily a much more basic world, one outside cognitive consciousness. If we can experience for a moment that reunion with chairos, we can know oneness. We were all gods then; and as Moses, Jesus, Muhammed, Buddha, Mahavira, Manu, Lao Tzu, Confucius, and all the great teachers have said, we can be just as holy now. If we can recapture that fearless, innocent perspective, we can find ourselves again at any age.

To accomplish this, we have to pre-stimulate more brain mass, and the repetitive physical activities need not initially be meaningful. In fact, it is often better if the practices at first seem meaningless. They will, through repetition alone, become meaningful in and of themselves, as described in Chapter 7. Most religious practices contain a number of regular rituals which by their nature are carefully repetitive. If enough of the brain is eventually associated into interconnected networks during religious rituals, a hormonal surge could cause the entire cognitive mind to short out. This could produce everything from out-of-body experiences to rapturous experiences, from dreamlike scenarios to superhuman exertions.

The aspect of neural activity that makes this sort of event possible is the ability to exhaust layers of consciousness by repeating the same stimulation, just as vision will waver and blur after staring at the same object for too long. If the brain is prepared by a combination of multiple repetitions in enough locations, like the clogged Internet server besieged with a swarm of demands, the energy in a large, synthetic associative network may build up high enough that just a bit more stress can collapse major parts of the normal thinking process. The neurons just stop firing for a moment to get some rest. If this happens, the informational overload may be automatically shunted downward to a more basic level acting as a fail-safe, which itself may have been similarly over stimulated. Like a domino effect, this spreading cascade alerts the adrenal cortex, and if the resulting endocrine surge exhausts the limbic system as well, our entire sense of self can melt momentarily in a powerful experience analogous to the feeling of being dissolved into an eternal moment.

The experience is often unexpected, but getting there requires mindful and attentive physical action that is once again repeated, only this time not in furtherance of any personal goals. It is ritualistic and recognized as such. The fingers move the beads for the devotees of Mary, Krishna, and the Buddha; the mind directs faithful hands to fold and pray; the Muslim bows to Mecca in a precise pattern, hands out, hands down. There may be familiar music. Familiar scents, such as incense, will excite the olfactory cortex. There may be ceremonial swaying, walking kneeling, ritual hand movements, ritual dance, or prostrations. Shakers danced in circles, Baptists clap, lamas move their hands in precise ritual gestures, Hassidim daven, Sufis spin, Chinese perform T'ai Chi, priests raise the Host.

Now we add other subroutines. To prepare the auditory cortex, we repeat a familiar line of syllables in our mind over and over. It can be a prayer, it can be a mantra, it can be a hope or a dream. We repeat it until it comes without any thought at all. Eyes closed or fixed on the altar, the image, the candle, or even the mental picture we have memorized, we keep our visual cortex in a similar state of synthetic repetitive saturation. To engage the prefrontal lobes, our future forecasters and time sequencers, we project the same expectation. All of these we repeat together, over and over again. Eventually we will have created such a huge network of memory and repetition that, if we're overexcited or unusually stressed, when the hormones hit, the higher brain could go up like a munitions dump, dissolving ego into whatever our culture has taught us to call the universal timeless experience of fulfillment and love. That is, if we have a faith. Otherwise, the experience might be entirely disorienting.

Once again, we are using electrochemistry to lift the roof off its lintels, but it is the only way that we can become oracles ourselves. It can easily leave the unprepared babbling in ecstatic syllables until normal processing is restored. By synchronizing the right physical and mental practices, we can greet God, ally with Allah, rally with Ram, dissolve in the Dharma, and come back blessed. Real spiritual masters and saints know how to do this from a standing start, but they have been working at it a lot longer than most of us. Any apparent shortcuts, either through compulsive religious activity or overuse of the powerful mental and physical practices of Asian meditative and tantric traditions, can actually be harmful to the unprepared, leading not only to hurt feelings and headaches but mild mental derangement as well.

Luckily, there are so many forms of gentler activity to engage in, both with others and by ourselves, that our personal meditative or devotional practices can easily be integrated into our lives. In time, our keys to the good times become easier to find as we discover sincere ways to lose ourselves not out of our minds, but very deeply into our minds. There we will find the answers we were looking for, and often when we least expect them. It is interesting to speculate when, and how, the powerful effects of complex interlocking repetitious physical and mental activity could have triggered the first transcendental experience. One possible scenario could have been the shuffling "dance" our early ancestors would have performed around the guard fire in front of the hunting camp or lean-to.

It is very, very late at night. The wind is arid and warm in the summer season when the grasses dry up, and the small bands of early humans must move from place to place seeking water and scavenging food. We are back in the time before history, more than eighty thousand years ago. The brain is now large enough that our early ancestors are finally living as much by their wits as by their primitive weapons. Life can be harsh, the dry season especially so. The small "family" sleeps as the hunter is keeping watch. The darkness surrounds him with sounds and stirrings. The ability of the prefrontal cortex to form a future is improving, but his future doesn't look good.

Beyond the faint light of the flickering coals are the jackals and hyenas. They are hungrier than the hunter, and he knows it. The woman is asleep, his cousins as well, the infant is ill. The hunter must rise again and tend the fire. If it dies, if he falters, the animals will come. He knows that too. Alone, or with a brother or clan cousin, night after night he shuffles about the embers, waving his throwing-stick, shouting hoarsely into the darkness where eyes lie waiting. Hour after hour it continues; the coals glowing at the center of his exhausted circle, the waving stick, the waves of memories and hopes, always the same, echoing through the auditory cortex in unspoken supplication.

"Come dawn, come morning light, come before I fall asleep, save me from this night of darkness, this night of fear." It may not have been spoken aloud, it may not have been in words, but it was the seed of what would become a chant or a prayer. All night the endless circling, the same movements, the smoky smell of the fire exciting the olfactory cortex and priming the hippocampus, the same words, the same worried fears exhausting one layer of cells after another, the visual cortex fixed on the glowing fire against the blackness of night, blurring, circling. His eyes grow bleary; his droning chant continues as he beseeches the sun to rise.

How many nights has he shuffled in that circle? Nearly every night as they camped across the dry African savannah. Every night the same ritual is repeated, the same endless dance, the same exhausted prayer. The combinations have all been there: rhythmic repeated movement of a watchful nature, neural exhaustion at the perceptive level, visual images, the smoke, an exhausted limbic system and a mind filled with feelings of loss and longing. It took only two days of sleeplessness for Charles Lindbergh to see angels in his cockpit. We can exceed our limits for only so long.

It was probably the same for Moses, exhausted in exile, staring into a desert sunset in the Sinai, and for Saul, about to be Paul, swaying rhythmically under a hot sun on the back of a donkey carrying him to Damascus. Gautama sat famished and troubled at the base of a tree as Sujata approached; Muhammed listened, alone in a cave in the desert hills as the angel spoke, Jesus paced and prayed for forty days in the desert. Saviors and prophets, intense, searching, all looking for their inner light. The truth speaks out of the sky, from blazing bushes, from angels, maidens and Satan himself, and it can be pretty violent the first time. It knocked Paul right off his donkey, converting him on the spot. Moses exited to start the Exodus; Buddha found the Four Noble Truths; Muhammed wrote the Koran; and Mirabai and Sri Chaitanya danced in joy from town to town.

Back in the endless rhythms of a prehistoric night, our ancient ancestor is nearly into an auto-hypnotic state. His movements are on autopilot, his exhausted consciousness at the sleep threshold, his eyes barely open. Sleep tugs at his mind. He falters, and the throwing stick clatters onto the rocks. He lunges forward, skips a beat, trips and stumbles toward the fire. He jerks back, the flames leap, a jackal howls a dozen feet behind him, and suddenly it's just too much.

The howl reaches the auditory cortex, but the associative networks are starting to unravel in overload. Signals break down as swarms of neurotransmitters clog uptake slots, but there's not enough room. It's simply too much to handle. In waves, unleashed chaos begins to surge through the overloaded neural channels, collapsing levels of sensibility like floors in a building and plunging the hunter into the fail-safe of sheer being. As his cognitive world dissolves into uncontrolled neural saturation, consciousness veers into pure asynchronous reality without any limbic limitations.

He staggers, momentarily stunned, and drops heavily to his knees before the fire. Is this death? The world sways and sparkles; he reaches out towards his sleeping mate and child. His companions, awakened now, see it all. What is he doing at the fire? Inside his bowed head, the whole history of his labors and devotions is avalanching into the present moment, and his body swims in hormonal shock. The overstimulated and exhausted cerebral lobes blank out, the cerebellum seizes, the prefrontal lobes wail a chorus to the brain stem, the hippocampus joins in, and the limbic system disinhibits. An adrenal rush floods his veins just as reality suddenly unhinges.

Anything is real now, and anything can be real. The hunter jerks upright like a puppet on a string. The adrenaline surge hits the visual cortex. His vision clears, his movements are sure and confident, and his limbs glow with inner fire. His prayers are answered: God has just kissed him on the top of his brain, and he knows that he is the one and only. More than that - he is strong, and he is chosen! Heart pounding, he strides in slow motion to where he keeps his stone axe, grabs it like a toy, and, screaming like a demon, dashes into the night, smashing jackals into jackal chops.

The first time was unrecorded, but it happened; and when they came to him the next day with the gifts and the fearful respect, there was something new on earth. He had discovered the first direct connection to something beyond ourselves, something that made us much more than ourselves. We, in turn, had discovered the first holy man. There would be many more.

The Varieties of Grace

If our Paleolithic hero learned to go through all those various preliminaries the same way again, if he did it enough times, or added any intoxicants for a little booster, he might have become the first shaman. Over many millennia, the fire and the steps became stylized, weapons and implements became sacred objects, and heartfelt utterances were formalized into chants and prayers. The holy wisdom took words, and was made poetic, but it was always the same eloquent expression of the human spirit articulated by those gifted with a perspective beyond the virtual reality of personal time and space.

Over time, the ability to combine activities that could create an interlinked symphony of neural excess that may be the catalyst for spiritual experience became a secret understanding, unspoken and acknowledged with great difficulty by those who were touched by it. It has not been, nor will it ever be adequately described because it is a temporary brush with the mind we knew before we knew speech. In fact, the experience often leaves us speechless. Nearly twenty years after this imaginary scenario was first portrayed in *Neurotheology*, the *National Geographic* seemed to confirm it. In December, 2014, in his epic series re-tracing the journey of the earliest humans, Paul Salopoek wrote: "The anthropologist Melvin Konner writes how the *num* masters of the !Kung San, the shamans of the Kalahari – members of perhaps the oldest population on the planet – induce a spiritual trance through hours of dancing around campfires. Such arduous rituals deliver up to 60,000 rhythmic jolts – the number of footfalls in a days trekking – to the base of the skull. The result, Konner says, is a psychological state that we have been questing for since our species first dawn, "that oceanic feeling of oneness with the world."" When our mental systems are momentarily forced beyond their limits, we may finally experience the limitless. Going beyond ourselves we gain an entirely different perspective, and may find our true place in the world we usually take for granted, a world distracted by the routines we repeat to comfort our fears when there is such power available to us all just waiting to be claimed and experienced.

The many paths to a renewed vision eventually became bound into religious and mystic traditions wherever human culture developed, embodied in innumerable cultural variants wherever there was a priesthood and a tradition. We are actually familiar with many forms of mental and emotional self-cleansing, experiences which may provide some with insight, others with wisdom, and all with a richer experience of life. All religions, cults, and even newer holistic philosophies have these practices available within them. For any Christian seeking a stronger faith there are polite prayer groups or passionate Pentecostal preachers to raise the spirit. The Orthodox Jew sways at shul, his kids kibitz with the Kabbala. Muslims worship five times a day, follow Sufi saints, or engage in Shiite S&M, smiting themselves with stones in a Farsi frenzy. Hindus and Buddhists have a particularly rich collection of meditative and tantric practices that serve to manipulate and massage the mind in precise degrees. Depending on how far we wish to take our involvement, we can self-generate everything from the warm glow of compassionate fellowship with co-religionists to the nearly uncontrollable full adrenaline surges found in ecstatic singing, talking, and dancing.

We can, over time, learn to let go of the ego just enough to work a little better with our friends, or go all the way and dissolve our personality into the mind of the universe. The chant can be "Hail Mary," "Allahu Akbar," "Baruch Atoh Adonoi," "Om Mane Padme Hum," "Nam-yo-ho, Ren-ge-kyo," or "Hare Ram"; and there are many more. They all work equally well, so say "Hallelujah," "Thank you, Jesus," and "Amen." If we wanted to take the trouble, and some do, any of us could devise personal movements, mantras, prayers, and rituals. With enough practice, they would probably do just as well.

On the other hand, it's easier on the emotional traditions of our limbic system to use a method that is historically familiar and natural to us. If we are deeply faithful to one religion or another, or even just make use of helpful devotional or meditative practices, it is easier to find inspiration within our own culture. Nearly anyone can follow Krishna with enough devotion and bhakti yoga, but getting back to Jesus may be easier for a lapsed Christian who was once a child who loved Bible stories.

Once we are familiar with some of the neurological staging behind our transcendental forms of pleasure seeking, it should come as no surprise that none of this can be accomplished without patient and sincere practice. We can't rush biochemistry, nor can we ever predict when we will harness enough of our mental energy to dissolve chronology and share the incomparable experience. As Diana Ross put it so simply, "You can't hurry love; no, you just have to wait." This is probably why both St. Paul and John Calvin stressed that one cannot achieve heaven by doing good deeds once in a while, and why Buddhist tradition insists on lifetimes of mindful practice. To even start our journey to our fulfillment we must be focused on our daily life and the pleasures we find on a moment-to-moment basis. Our attention must be in the present tense, not wandering about in our past or lost in expectations of some future fulfillment. As it happens, humans are eminently trainable in this type of self-improvement. Real blessings and a growing enlightenment can arrive in less than a few years, often when we least expect it.

If we prepare ourselves and practice diligently, fulfillment will come looking for anyone who is truly ready to accept it; and it inevitably arrives in the present tense. When traditional prayers or practices are done with any consistency, there will always come a time when the practitioner will begin to notice that the world is, for some reason, looking better and more inviting. If we feed the mind a more balanced natural diet, each well-experienced day filed away creates expectations of a similar future. We can't change our chaotically woven system of human consciousness, but we can load up the loom with good times and start to watch the patterns change. We can even darn up the holes in the networks we don't like and fine tune our virtual reality to the tunes we enjoy.

The whole purpose, of course, is to reach the point in our life where, from a Western religious perspective the Kingdom of God is at hand or when, from an Eastern perspective, we are living in the Dharma or the Tao, on our true and natural path. From a systems perspective, we would say that we acquired and uploaded some debugging apps to override some glitches in the original software, letting us reset our goals and restart our life. In reviving and maintaining a full and active involvement in the world around us, we are freed from the habitual gridlock of virtual ego and returned to an exciting, moving world with a peaceful, personal center.

Those drawn to complex ceremony and ritual rarely have time to reach out to others who do not dance in the same circle or chant the same prayers to the same God or spiritual guide. The sincere and simple paths to empowerment are there to lift us out of both past and future and rededicate us again to the present - the only place shared by us all. In finding ourselves again we are not confined to co-practitioners, we are liberated to go and involve ourselves even more fully in the world around us. The big secret, if there ever was one, is that we each have within us the ability to do it all by ourselves. There is a good reason that Jesus directed his followers to pray in the privacy of their rooms and why the Buddha directed his students to find calm and quiet places for meditation. The practices which reach the deepest are truly self-tailored. They are not group events, they are personal, and they are precious.

Finding lasting fulfillment for a human, then, is nearly the opposite of the rat with its pleasure button. While animals drive themselves quickly to exhaustion, only a small percentage of humans are that compulsive. In fact, most of us do learn to enjoy the conscious generation of joy and happiness as a regular experience that makes life itself more vibrant. We can get excited about our art, our craft, our dance, our friends, our family, our skills, and even our mystical mental spirituality. We can all learn to live inside and outside our limits, learning about ourselves daily in a full involvement with the life we are actually living.

It's true that along with all the advantages of a fully evolved human consciousness we still experience unique problems. We slip easily into so many forms of mental mind-block, re-cycling, and repetition, not to mention finding ourselves caught in the gridlock of chronological time. On the other hand, it seems that we have developed some impressive mental retrofits and even specific applications to enhance and improve our own consciousness. This is what we've always been told, but we never really thought of it that way.

Applications? We call them "Loving," "Giving," and "Forgiving". Loving and giving keep us open and aware to the best life has to offer, leading us each into our better future. We all have enough love within us that we need never go searching. Forgiving is the permanent delete that patches the past. Prophets and teachers have been telling us this forever. Putting these words into practice, we see why the great religious leaders had so little to say about ritual. Jesus never mentions speaking in tongues, nor did Muhammed do Sufi dances. The Buddha walked with his monks teaching from town to town; he did not sit Zazen in a cave. Our guides were not telling us to give up on life, they were urging us to give ourselves out to life and to enjoy the greater community of the entire human family. This is why they speak not of power and wealth, but of simple consideration, forgiveness, generosity, and above all, compassion, love, and kindness.

These are the sorts of pleasures that only humans know anything about, and when we practice them, they bring us closer to the best of our own humanity. As natural pleasure seekers, then, we have so many ways that we can travel. There is everything from full-blown fantasy to real self-discovery and all the stages in between. We can be angry at life's obstacles and get our excitement stressfully, or we can be enthusiastic about life's challenges and get it cheerfully. We can all find ways to make life as stimulating as we want it to be, we generally use the ways we've become accustomed to, and most of the time it works. But the careful and graduated steps to inner tranquility and personal fulfillment are also there and available to all with the will to improve and the patience to keep at it. Day by day, step by step, and moment by moment, we can update the tapestry of our own virtual perception with mindful attention to our craft, our art, our hopes, our practice, and our prayers.

With these to help us, we can re-pattern our mind for easy gladness, and make personal happiness into a habit. Our tools may be molecules, and heavenly experiences are still dependent on hormonal states, but we can get an entirely new outlook if we give it a try. This is one world that gives us more chances for a good time than we ever thought possible. It is, after all home of the Big Apple, Fat City, the Garden of Eden, the Kingdom of God, and we are right smack in the middle of it. In going beyond our personal, cultural, and even our conscious limitations, we have a chance to awaken to a better world that was always there for the taking, and the vision will change us forever. With a fully evolved human mind and body, we can have a better time of our lives than any other creature on this earth.

Sadly, our recent record in this area hasn't been very good. As a species, we haven't been able to do a lot of damage for very long, and yet for most of that time we've practiced exploitation, destruction, and extinction. At the same time, we have the best opportunity that we have ever had to start acting like humans for once. At the moment, our greatest challenge is saving the earth itself from the mistakes we have made. If we can cooperate in this extraordinary work and return our world to the state it was when we woke up and realized we were here, it would be the greatest monument of all, the sort of undertaking that only a united human mission could accomplish. We could all be missionaries of that faith. Now that we have available the tools to organize and work together, we can get something done. The Tower of Babel is prehistory. Can we talk?

With a task like this, for all the beauty and poetry of the multiple languages of our planet, it is perhaps our greatest blessing that we share and accept the common language of science. We can communicate now and collaborate, and we surely will. In the inevitable process of mixing and meeting for common causes, strangers join together to make it happen just as strangers once joined together to make each of us happen. All we need is love, it seems, and we'll always come up with something new. In the process, it seems inevitable that before long, those similarities common to our world faiths will become ever more obvious as each of us awakens to this uncommon consciousness we share.

We each own our own domain, our personal virtual reality, but we can share our unique visions and dreams with so many others in the only universe we know. We are the only species inhabiting six billion different worlds, each with a singular place in our hearts that we call home, and each pondering the question we all know. Will we get home in time? Would anything as magnificent as the greatest show on earth just disappear? Does it ever end, or do we simply fold the tents and take the circus with us as we start the journey to our next destination? What really happens when we die?

10

Soul Survivors

What Really Happens When We Die

> *"No one wants to die. Even people who want to go to heaven don't want to die to get there. And yet death is the destination we all share. No one has ever escaped it. And that is as it should be, because death is very likely the single best invention of life."*
>
> *– Steve Jobs (1955-2011)*

Life is a lifetime falling into death. From birth we trace an arc, tossed up into the living for a time, but even as we are loosed into life, our destiny is determined. Life, it seems, has a catch to it. There is an end to it. Eventually we must touch down, and we hope the catch is gentle.

As children, we think nothing of it, too taken with our vital present to imagine a finite future; but as we grow older, we begin to notice mortality, and before long we learn we are not everlasting. Long before we fully comprehend the certainty of our temporary existence, we pray that God will take us into heaven when we die. Wherever that is. Whenever that is. As adults we interpret it to the young, trying to explain the reasons behind the experience of life, but when it comes to something as common as death, we are still like children. We know where Santa Claus gets the toys and where the Easter Bunny gets the baskets, but most of us are still hopefully expecting that if we die before we wake, we pray the Lord our soul to take. Even the most rational among us usually agree that when it comes to that inevitable, ultimate, and final transition, God only knows what happens then.

For many more, avoidance is the best refuge against a disturbing realization. As the world population grows, we notice more and more people dying all the time. It seems to be trending. In response, we keep our minds fixed on the here and now, rather than on the where and when. We all live our lives, and then?

And then ... will the heavenly odds-maker collect the bets and will the first person to the other side please tell us what happened? Am I Brahman, a spirit, or an angel? Did Jesus love me, or did I miss Nirvana and I'm about to be recycled as a turtle for some Buddhist sin? Are these the Elysian Fields, the Happy Hunting Grounds or, wrong turn, doggy heaven for Rover? The more we think about it, the more we realize how undefined this most inevitable of destinations remains. We know more about the moon than about the experience of death, and very few have gone to the moon. Those who went, however, returned, and that is the big difference. The moon is a temporary destination; death is always forever.

Where, or how, we spend that forever remains for too many a bothersome unanswered question. Even the most religious souls are curious; this is an innocence nearly all of us keep throughout life with much guesswork and very few authorities. We cannot speak with authority ourselves, and those with real expertise have nothing to say at all. Dead men tell no tales. And so, preferring something to nothing, many accept the various descriptions of everlasting life, or lives, as handed down by our traditional religious or spiritual beliefs. Those who have found a path they can trust know the peace of the mighty and the comfort of the meek. As we grow older, we begin to understand that we all want that assurance. If truth be told, nearly all the non-believers would love to have a reason to believe.

Guidebooks To Forever

Comparing the afterlives described by the world's great religions, we begin to notice similarities. At first, singing in angelic choirs doesn't seem quite like getting off Buddha's wheel of life, but there is always ultimate peace. It is always a blessed journey or a return to a higher and better place where the woes of earthly life are left behind as we take up a new existence without suffering or pain in a world without end. Living in the cross-cultural currents of our global society, we sometimes have trouble reminding ourselves that less than a hundred years ago, wherever we were, we were either a believer or an infidel. Today, although fundamentalist sects of major world religions still bar non-believers from heaven, most thinking people would agree that Gandhi was working on the same wavelength as Mother Teresa and would allow for cultural variations. This was unthinkable a century ago, when major world religions were more geographically centered. But what about death and beyond?

Most religions seem to have reached a general consensus that there is a series of stages. At some point of time between when we stop breathing and when we start coming apart, the non-physical part of us (soul, mind, spirit, atman) takes a journey to another place. The mortal body, which was created at the same time as the eternal part or which houses it during this life, usually proceeds to compost. However, the soul, spirit, atman,etc., continues to exist in a mindful, if disembodied, fashion as it starts its journey onward. Often guided by a heavenly light, we are transformed and welcomed to our eternal home.

There may be an initial purging, depending on what we did during our life on earth. The Purgatory of Roman Catholicism bears similarities to the Tibetan's frightening Bar-do world between lives. Swedenborgian theology describes a time in a spirit world that makes us fit to meet God. It's important to note that we cannot stay in these places indefinitely. Whether pausing in a purgatory or detouring through a few extra lives to clean out the karma, sooner or later we progress. This is all for ordinary people of course; true saints go directly to the good place and real evildoers go straight to the bad place.

Then comes judgment. Our deeds are totaled, our purgations accounted for, and we are assigned to a far longer stay somewhere else. If we are now acceptable, we go to heaven, Brahman, Nirvana, or the Pure Lands, and stay there forever. If not, it's back to purging, more lives, or worse. Most of us eventually get to a nice eternity, which comes in nearly every variety depending on the time, culture, and nature of the scripture. It's always the destination to die for, but these days, with so many world faiths, our heavenly destination comes in many forms.

If the GPS is off, however, we might have a problem. A Catholic who had made it to the most common heaven of Japanese Buddhists, rebirth in the Pure Lands, would probably assume that Franciscans had charge of eternity in this merciful agrarian paradise. The displeasure of the Viking waking up in Jewish Sheol, a very sober destination, when he expected the eternal fraternity party of Valhalla, is not recorded in language we can repeat here. Mormons enshrine marriage on earth, so in Mormon heaven you keep your mate, but a blessed Muslim is free to meet new partners in Islamic bliss. Serious Christians become joyous celibates while celibate old Himalayan monks could find themselves manifested as minor tantric deities in eternal sexual union with the appropriate consort, complete with four arms, prayer beads, and a yak-tail fly whisk.

Hell, likewise, seems to vary to the extent one accepts a literal interpretation of the Holy Word. In the hot lands of the Middle East, birthplace of Judaism, Christianity, and Islam, cool is heavenly, so hell is hotter than blazes. Ironically, "Hell" is the name of the frigid Norse underworld, closer to Dante's ninth circle: a frozen lake of ice. They would have loved some heat in original Hell, where frost giants stalked and cold was the killer. Buddhist scriptures describe both hot and cold hells, further subdivided by Tibetans into picturesque categories and names such as *a-choo*, a sort of endless cold in the nose. This may be why orthodox Taoists borrow Buddhist heavens but choose Taoist hells. Jains have the most hells, exactly 8.4 million, although it seems to have escaped notice that the Muslim Jahannam appears to be the same place as the alternative Hebrew hell, Gehenna - a truly hellish prospect for any evil Arab.

The reason that hell is still not the final judgment is that most, if not all, hells appear to have a back door. An abjectly bad Buddhist will simply be recycled in rebirth after rebirth until his karma is all gone, no matter how long it takes. Christians have until the very last moment to make peace with God. Even if the evil unfortunate ends up in the place with the pitchforks, the message of Christ promises forgiveness whenever true repentance appears. This "get out of hell card" last-chance option has been official for Catholics since the Second Vatican Council, but it's been Hindu faith forever. Even the most sinful swami could return only so many times as a street beggar before eventually getting back on the straight track to Shiva. The trip is just longer and rougher for some than it is for others.

Since most, if not all, unpleasant detours are apparently incurred or avoided by the manner in which we live this present life, the great religious teachers provide insights and methods not only useful to us now, but also able to transport us to a good eternity without too many intermediate stops in those unfortunate places. In fact, one of the major ways we can distinguish one religion from another is by its afterlife. Our great saints speak similar wisdom in differing tongues, but the Pope and the Dalai Lama have distinctly different retirement homes when their good works are done. The reward of Christianity is instant heaven, while the enlightened bodhisattva returns for future lives spent helping others. This is eternal, either way.

Once the journey is over, we spend the rest of forever in the nicest place imaginable, either that or in endless relief that it's over. Theologians write of everlasting oneness; those with more vivid imaginations have for centuries expounded on the unspeakable, ineffable, final fulfillment of our last stop. It's always our eternal home, always just what we wanted. Like happy, heaven seems the same to all peoples and is still very personal to each of us. Death, not life, appears to be the force that unites us all, and yet it still separates us. We all die and most hope to journey on, but regional religions still tend to determine our personal beliefs and faithful expectations. Those without faith suspect that such expectations are pure romance, but most still hope to find the place our religious faith has promised.

The big difference these days is that most of us are now aware that the world population faces a score of destinations. Our explanations may come in many languages, but they all seem to describe the same experience. Different guides report scenery appropriate to their custom and culture, but all cover the same territory and come to the same place when earthly time stops and eternity begins. The promise is always fulfilled; by God, by Allah, by Dharma, or by the Tao. The steps are so regular and consecutive as to suggest a common heavenly blueprint. Might there be a basic, underlying, universal pathway to the beyond?

Again the possibility arises: Could this be neurological phenomena? So far, various aspects of the neural process itself appear to be responsible for a number of the unanswered questions we all face. Death must be the granddaddy of them all, the biggest question in the mind of mankind. The late theologian Paul Tillich named the three greatest fears of man: death, madness, and the life lived in vain. Can that first fear be conquered? Is it crazy to even try? Most would agree that any attempt would be in exercise in vanity.

The main problem is that although we are all promised appropriate afterlives, no scripture explains just how we shift over to this timeless universe that seems to appear only when we are dead and gone. None of them come with a shop manual to describe how we can accomplish this leap to immortality given the only tools at hand: our old, sick, dying mortal selves. These days, most of us don't like to believe in magic. If it takes a mystery or a miracle to get us to heaven, that seems a little awkward. If it's really possible, perhaps it's time we came up with an explanation that makes sense. Now that modern medical technology seems able to keep any of us, or for that matter any part of us, alive almost indefinitely with various implants and devices, there is a renewed interest in just what happens afterward. No one in recent history has died and returned to life, and nobody yet has been known to survive brain death. We do, however, have volumes of reports from those who got close enough to stick a toe across and beat it back before it was too late. By reviewing available information from these near-death experiences, often referred to as NDE's, we begin to get a picture that may help guide us toward the explanation we seek.

There are many common themes: a miraculous transformation, departure from the physical body, heavenly beings, often a white light or passage, and the sense of timelessness. Only the details seem to be cultural. Nirvana never arrives for a devout Dominican nun. Holy Hindus drop their bodies and achieve samadhi, but they never meet Mother Mary. It is our own life that we re-experience, our own relatives who greet us along the way. Left unanswered is how we can greet our grandparents if they are off with their own grandparents: the paradox of the infant grannies. Holy books seem strangely incomplete; the inevitable crowds of Chinese in paradise are simply not mentioned in any Christian biblical text. Even heavenly angels meet cultural expectations; winged for Christian, non-winged for Hindus. Heaven is always a curious combination of human universals and cultural specifics.

Almost all of us will lapse into brain coma before we die, but each year a few make it back to describe the experiences they had. It seems a new "visit to heaven" book climbs the best-seller lists nearly every year, the ones written by physicians or young people leading in popularity. Although many are beautiful and poetic, one stands out.

On the morning of December 10, 1996, Jill Bolte Taylor, a thirty-seven-year-old Harvard-trained neuroscientist, suffered a massive stroke. A neuroanatomist by profession, she observed her own mind deteriorate completely to the point that she could not walk, talk, read, write, or recall any of her life, all within the space of four hours. As the massive neurological disaster unraveled her higher cognitive centers, all her rational and time-oriented functions became unavailable. In her 2008 book, *My Stroke of Insight,* followed by an easily available TED talk viewed by millions on YouTube.com, she describes in vivid detail the complete disruption of her universe as her virtual reality crashed around her.

Taylor found herself alternating between two distinct and opposite realities: a euphoric retreat into the intuitive brain, a realm of complete well-being and peace, and the intrusive attempts of the logical, cognitive centers which quickly recognized she was having a stroke and doggedly enabled her to seek help before she was lost completely. Her recovery took six years, but today Taylor is convinced that the stroke was the best thing that could have happened to her. It taught her, she says, that Nirvana is never more than a thought away. Her widely acclaimed work has become a source of comfort and inspiration to countless stroke victims and their caregivers.

In a classic 1980 study, a number of survivors of near-death experiences were cataloged for similarities by Dr. Kenneth Ring, one of the first physicians to conduct serious research into these phenomena. Placed in the order they were perceived, these reports suggest a series of common experiences. Subjects reported "peace and contentment" (60%), "detachment from the physical body" (37%), "entering the darkness" (23%), "seeing the light" (16%), and "entering the light" (11%). Since most patients who suffer the sort of trauma experienced by these individuals do not recover, survival rates would naturally favor those who experienced only the first stages of brain coma. Those that report "entering the light" account for a small percentage, probably because most of those who get that far don't come back.

Along with near-death revival stories, there are the last words of those who died describing their final visions, often leaving poignant images of a place beyond. Interestingly, these visions are almost uniformly pleasant and often include visions of parents or other relatives and friends who had died before. Krista Gorman, who survived eight minutes of clinical death while giving birth, describes a beautiful Eden—like world filled with love to Morgan Freeman in a *National Geographic* segment also available at YouTube.com. The poet and religious mystic William Blake, present at the death of his beloved brother, recounted seeing his brother's released spirit ascend heavenward "clapping its hands for joy." Steve Jobs' last hours, steeped in his yoga practice and expected for years, were determined and mindful as he said his last goodbyes. As his sister Mona Simpson recounted in the *New*

York Times, "Death didn't happen to Steve, he achieved it." After making it through one final night, wrote Simpson, her brother began to slip away. "His breath indicated an arduous journey, some steep path, altitude. He seemed to be climbing. But with that will, that work ethic, that strength, there was also sweet Steve's capacity for wonderment, the artist's belief in the ideal, the still more beautiful later." His final words were monosyllables, repeated three times. At last, surrounded by his family, he gazed at each in turn and, looking up and beyond them he peacefully surrendered. His last words were those of nearly speechless awe. "Oh wow! Oh wow! Oh wow!"

All human cultures have religions, so the possibility arises that the heavenly images and experiences common to all religions might be common to human consciousness itself. Since similar images appear in the words of revered prophets and sacred texts as well as first-hand reports of near-death, or clinical death and revival, it suggests there might be a neurological explanation, something common to all of us, that we could all experience in one form or another. If the developing brain left us with a taste of eternity as we matured into our human consciousness, it should be able to bring us back to that meeting place before we leave. Saviors and prophets have always been able to tell us where we went after death; it seems science may finally be ready to provide a reasonable explanation of how we get there.

Welcome Home: Return To Eternity

So what happens at death? The nature of the experience of death may already be apparent to some. If we once spent forever winding up the mental clock that ticks us through time and space during those endless eons between conception and age three, it will take just as long to wind it down. The human brain at the point of death has over eighty billion fully functional neurons. Each one is different, each is alive. As death arrives, they cannot all suddenly leap up and die at the same moment; that would be impossible. They must die off over some period of time, and their more vulnerable functions would fail first. From the most sensitive dendrites on the most exposed cortical cells to the most embedded neurons in the brainstem, the brain dies by degrees.

Since it is the activity of the human brain that creates, permits and limits our awareness of anything else, how will our awareness change as the brain changes during the time of death? Unless the brain is suddenly destroyed, the stages of brain death cannot vary much from one person to another. We have known many forms of consciousness since our unborn days when our brain was a fraction of its current mass or complexity. We're bound to lose our more recent mental capabilities long before we reach any final end. Investigating this question, one of the most important advantages made possible by brain imaging has been the ability to observe the manner in which brain activity degrades and diminishes by stages in a regular sequence. From a strictly medical point of view, the brain will begin to sustain irreparable damage at normal temperature after ten minutes without oxygen. This reinforces the conclusion that during the process of normal human brain death, the major biological supports of consciousness couldn't instantly collapse all at once. In other words, even if we wanted to, we couldn't just pull the plug on consciousness. It must simplify in a somewhat predictable progression.

As we start to die, we fall into irreversible brain coma. Brain coma, however, is not by any means the end. It is the end of this worldly consciousness, but also a return to an earlier form of consciousness and an earlier universe. Recent MRI scans detailing the sequential stages of unconsciousness have identified the recently evolved prefrontal cortex as the first area of the brain to fail. Making chronological time is a delicate operation. We've all experienced those moments when time stood still, we have all had dreams that seemed to last ages only to waken and discover we had been asleep only a few minutes. When we are asleep and the time sequencing system in the prefrontal lobes goes off-line, we know that dreams can pack months into moments. But where do our dreams come from?

Harvard researcher Alan Hobson believes that dreams are neither Freudian films nor mystical guides, but artifacts created by stimulating the higher brain centers with irregular bursts of neural static from the brain stem during sleep. Emotional states aroused in this manner take visual form, but time and abstraction are off-line. A dream starts in now and ends in now, and there's no reflection or deep thought. This is a small example of how an unconscious mind can fully experience a consciousness with its chronology unattached to worldly perception. Since we dream using images synthesized from memories of people and events in our own life, our dreams are completely realistic.

No matter what our cause of death, then, waking consciousness must go unconscious before death comes. This means that in most instances we could still be aware, unconscious but in a dream-like state, even as the brain dies; cells winking out at random, axons sending their final messages, dendrites reacting, failing, and at last falling silent. The experience of the simplification of our brain would likely be perceived as the gradual simplification of our mind over a period of time which could seem endless. As the neural nets unravel, we will gradually return to the timeless eternity of the undifferentiated mind that we knew since we were created, and long before we were born.

The progressive stages of brain death specifically responsible for the basic near-death sequence as reported by Ring's subjects have been known for some time. They were collected and put into general order some years ago by Canadian neurosurgeon Leslie Ivan. The brain starts to die as the delicate balance of its blood biochemistry begins to change. Usually, something interferes with oxygenation, and as the oxygen levels drop, the neural firing rate begins to decrease. This creates the pleasant, dreamy "peace and contentment" felt by so many who are near death. The buildup of carbon dioxide and other toxins in the blood now starts to create distortions in cortical firing patterns, while deep in the limbic system, specific endorphin receptors begin to react to the falling oxygen levels. Generalized physical sensations, dissociation from the body, and even euphoria begin to occur. At about the same time, the visual cortex begins to fail along with the chronological sequencing structures in the prefrontal cortex. Our sense of time and space begin to waver as we are bent gently back toward our beginnings. As memories and emotions are released from time, images from the past begin to flood a consciousness that is no longer either exact or discriminating. Soon, blood loss or changes in blood chemistry have progressed so far that cortical brain cells are beginning to die at random.

The visual cortex is a sensitive and sophisticated structure: it is relatively near the surface of the brain and therefore especially vulnerable. Visual memory patterns are losing definition and fade as neurons cells in the visual cortex begin to misfire and die. Finally, inhibitory rule structures crash, releasing the blend of all colors in the vision of a white light witnessed by so many. As the visual cortex continues to simplify, the color scale now begins to alter and dissolve back to the earliest color we knew, our primeval dull red, not the fires of hell, but the endless sunset that finally fades to the familiar darkness we knew from the very beginning, before our own dawning, before we were born.

The darkness now surrounds us, beyond the sunset, there remains a dull glow. Consciousness is quickly losing the last edge of specific definition as the continued destruction of the neural networks increases. The last fits and starts illuminate the great ocean of oneness with pinpoints of blazing energy, the stars guiding us into our new old universe. We are among the stars now, and we begin to move toward the distant light. That light is the very last signal, when others have faded into the gathering night. Like our last call from this earth, the dying reticular activating system, our old reality filter, surges, yanking consciousness tight for a final moment. We sign our names in this universe for the last time, and return to our final home. When we get there, we will enter the light and become the light. We know where we are going on our last journey because we came this way before. Now we return.

Each of us will, in time, one by one, join in this final shared experience. We must travel together with the mind that made us as the weave is gently unwoven. As our brain, the great analyzer and discriminator, moves moment by moment to the final and ultimate simplicity of one last cell, we are moving with it. At no time can we be aware that it is we that are simplifying; we no longer have a mind capable of discriminating thought. We are now moving backward into ancient memories we could never remember, woven in a simpler time. We are returning to the other universe we know, the universe we always knew, the one we have carried with us all our lives. Where do we make our transcendent ascension? Probably between our dissociation from the body and our arrival at the light, which is as far as has been reported. The specifics will always be personal, but the most detailed reports have come from some highly trained Tibetan lamas whose last words were characterized by careful descriptions of the stages in the dissolution of their worldly consciousness. The sequence, as described by the Venerable Lati Rinpoche and Professor Jeffrey Hopkins of the University of Virginia, even includes changes in the color of the sky as one begins the final journey onward into the Bar-do, the gone-beyond.

The dream sky at the beginning of death is initially bright white. White is the mixture of all colors at equal intensity, a good description of what the background would look like in early brain coma if the visual cortex had disinhibited. As time goes timeless, we lose visual definition, and the sky slowly fades to red, the lowest frequency color in our visual spectrum and the earliest color we knew. According to the Tibetans, we then see "points of light, like sparks." Finally, there comes darkness and the "setting face to face with the clear light of death." There is still a lot of brain remaining, but we are now as timeless and as sightless as we were in our seventh month in the womb. We are the one and only again, only this time forever.

Scientists prefer independent verification for theories that the mind may be experiencing these sequential distortions even as the brain is simplifying during death. In 2011, neuroscientist Antonio Damasio gave a TED talk on the nature of consciousness that has been viewed nearly a million and a half times. The sense of self, it seems, is very hard to lose, and persists all the way to the brain stem. Without a single scanner and with the most rudimentary knowledge of brain science, the lamas had been describing in detail the gradual death of the brain while the event was actually in progress. They never went beyond the "clear light of death" in their lucid descriptions; by that time they had stopped talking and "gone beyond" to the most profound and universal state of mind we will ever encounter in our lives. It is a return to our beginning. The circle is now complete, eternity to eternity, and all in one lifetime.

This does seem to be our path; but what would the personal experience be like? It would probably feel like a blessed event, as gently reassuring as our birth was once so bewildering - when was it, a few moments ago? With timelessness fast approaching, our lifetime will seem to have been but a short sojourn, almost a dream, in some recent world. As discrimination falters, we will begin again to remember forever, to see again the sights we saw when we had just arrived from where we are now returning. Tall beings, past lives, the rounds of judgment and rounds of forgiveness: the long forgotten past returns as time itself begins to stretch out, moment by moment.

Years appear now between the minutes of earthly time, centuries between seconds, eons between the tenths of seconds. Finally, as was promised by our God or our faith, we are returned to oneness forever, for had there ever been anything else? By this time we are timeless, as the heart, the mind, the soul, and the universe all merge in the journey back to one, the journey that will never end. Eternity arrives early. It comes with our final consciousness, and it comes for us just a few minutes before physical brain death.

We will never be able to perceive death itself; we will run out of time and self long before it gets to us. The final landing is gentle indeed; we have nothing at all to worry about. We all go home in the end to the timelessness of another universe that remembers nothing and is forever.

Although this description of the simplification of consciousness agrees with information we have from the scientific community, it must remain speculative. Final confirmation remains impossible because of the nature of life; there is a threshold below which a dying cell is dead and cannot be revived. The sense of self may persist all the way to the brain's most basic structures, but anything that would reduce consciousness to a universal state would probably kill off so many brain cells in the process that we might as well stay there. Further studies of people who have been revived from drowning or experienced near-death experiences have revealed that they actually suffered more damage from the sudden return of blood to the brain than from the initial anoxia. Dr. Taylor was actually very lucky; she worked hard to recover and there have been lasting effects. Most people, if revived, would likely suffer from extensive brain damage, remaining trapped in a body completely inappropriate to their mental state. When faced with the question of whether life support should be removed from the brain dead, Pope Pius XII suggested that, in irreversible coma, the soul might have already left the body. He was right, and in even suggesting it, he was demonstrating how easily religion can incorporate neurological perspective as a backup for wisdom that was always available. Still, the final proof will always be missing. Our best witnesses leave us before it's over.

Among the living, then, we can have no trustworthy reporters. Brains and minds in the process of development toward adult complexity are in the heads of people too young to speak and as yet unable to reason. Likewise, dying people end with their dying words unspoken; we never hear about their final destination. The Book of the Dead must still be read on faith, but hoping to attain eternity does not seem unreasonable. In fact, there seems no way to avoid it. As the result of time distortion, which must occur as we lose chronological controls, there is no way to know how long it takes to regress consciousness back to eternity. Brain cells are capable of firing hundreds of times per second; we could slow down to a graceful end in the blink of an eye. By the time we reach our own ancient universe, time effectively will have stopped for each of us.

Death is usually gentle, but even many violent deaths could not prevent consciousness from going out the slow way. There would be a swifter transfer to unconsciousness, but then our comfortable and steady return to eternity would begin. Unless the brain itself was severely lacerated, even victims of shootings, stabbings, car accidents, or massive loss of blood could proceed to their final journey without further pain or problems. Even decapitation wouldn't change the sequence; the shock would be traumatic, but neurons still take time to die. Common diseases such as cancer, heart attacks, or failures of major organ systems would seem nearly guaranteed to bring us to the same shore. Should a boulder drop on our head, however, there might be no death dream experience at all. We would be eternally in the moment before it happened because in a fraction of a second, all perception would suddenly disappear. Neural impulses travel at about eighty miles per hour; anything that smashes into us going faster than that would slam us into forever faster than we could realize what happened. We would remain forever in the moment before we never saw it coming. We wouldn't miss heaven; in fact we wouldn't miss anything at all. As long as we didn't see it coming.

This brings us to one variety of death that should at all cost be avoided: the violent destruction of the brain while in a disrupted mental state such as panic, pain, misery, or terror. In such an instance, regression could not occur and eternity would be the last consciousness available. Facing the gun that blows our head away, or frozen with terror in a damaged aircraft that has not gone instantly to pieces could be the worst death of all. For those unable to calm their minds when facing a sudden end, either through religious faith or powerful meditative ability, eternity is dismal; final terror and then nothing more. Every religion in the world has its ghosts. They are almost without exception described as the disconnected souls of those who died a violent or disjointed death. There was no last option. For those who ask "What about Hitler?" they have their answer: most of his victims went to heaven, but by shooting himself in the head, he probably made it impossible for himself. There can be no regression in a brain blown to pieces.

Karma and Compassion: Why it's Good to Be Good

There's a natural justice in the way it seems to work out. If our regression into timelessness is within our own mind, it is within a closed system. Any road to heaven must be paved with our own good intentions, or at least the memories of a life spent that way. At the start of death, we take leave of the open system, the world around us, and enter the closed system that exists only within us. Now our only reality is our virtual reality and we must reside in the world of images we made ourselves, our databank of personal recollection. Any recognizable heavenly or hellish scenarios will be mental constructions arising from our dying networks, just as we made our normal dream images from mixed-up memories when the forebrain took a nap and time was off. This time we will join with the dream. We shall not wake again to the world of troubles or pain; we are already long gone beyond that place.

From the neurotheological perspective, this one aspect of brain death provides the impetus for an ethical lifestyle just as convincingly as any religion currently practiced. It provides a rationale for living a good and decent life that makes better sense to some than any promise of rebirth, heavenly or otherwise. It has already been demonstrated that we cannot own universal reality in our minds, just our own virtual reality, our personal self-created neurological interpretive interface. At birth, and for the first three years, the plasticity of the growing brain prevents the organization and repetition that characterizes the mature thinking brain. It is only as we begin to create and extend larger networks that they become extensive enough to survive early disruption of brain function in death.

Infants who die in the womb, soon after birth, or even during infancy would naturally have a very easy return since there would be so little repeated experiential detail to disturb a smooth regression to our original mind. However, most of us, by the time we die, have a lifetime of memories to draw on. One thing is clear. The path we take during death may be outside the bounds of normal time and space, but it starts inside our heads only after our normal senses have shut down. The outside world ceases to exist, so we can live only among images from our own past personal experience. We are stopped at the point of death and sent packing to eternity with whatever we have put in our heads up to that time. This can be nice for those who have minds filled with memories rich in kindness and simple pleasures. If we spend a lot of time worrying or stressed, however, there could be roomfuls of blues that we might endure on our way to wherever after. In the first part of our death experience, everything will be happening at once and forever, but all the imagery must ultimately be derived from our own memories.

From this perspective, insofar as ensuring a pleasant "afterlife" is concerned, what actually happened at any point in our lives will never be as important as what our state of mind was at the time. If we spend our days being hopeful, helpful, kind, generous, supportive and reliable, we should have very nice "endless lifetimes" on the way to the final place. Fortunately, all sincere practitioners of world faiths have available to them a rich heritage of rules and suggestions for living and improving our life on earth. Since we pack our own bags for this trip, we should pack them with care. It will seem to be so much longer than our entire life on earth that we would probably do well to consider it every day.

This sort of death scenario completely supports the words of our saints and saviors when it comes to how to live life with respect to the afterlife. If we live by the words of Jesus, by the law of the Prophet, by the light of the Torah, or the systems of the Sutras, we should have very little to worry about later. If we make life hard for ourselves and others, we'll take the long and hard way home. It's simply inevitable, any way we look at it. This is also a good take on karma. The concept of karma in Asian philosophy speaks directly to the accumulation of neural nets that bias reality with personal attraction or fear and in this manner delay enlightenment.

Karma happens only with intention. Intention to do good produces "good karma" and intending to harm creates "bad karma." If we step on a bug that we hadn't noticed, there would be no karma, but if we set about looking for a bug to harass and step on, we'd create lots of bad karma. It was where our mind was at the time that made all the difference. Any conscious intention requires focusing the mind, creating the sort of memory we might encounter in death. The problem is that if the nets are unraveling, the memory might merge backward with the memory of another event, and we could find ourselves being stepped on by a huge bug in an extended time frame. Once neural destruction starts, entropy takes over, and God only knows what images will emerge as we experience the first stages of forever. This also supports the aspect of Buddhist philosophy which insists we cannot attain Nirvana until all our karma is exhausted. Whether one is going to spend those earlier endless lifetimes in wonderful places we created by a good and virtuous life or in various hells and scare shows derived from unpleasant memories, we cannot reach any final place until the brain has so simplified that our neural nets can't hold any images at all any longer. In fact, eternity with specifics may be just the first part of the experience.

This unusual time and space warp of "forever first and more later" is made possible because we are not using sequential time any more, unlocking the source of Bhairab's express in Varanasi as well as the source of Jesus's life everlasting. We can do eternity in a moment if we turn off time, but we still must reach that final union regressing along the path we made in this one-and-only life we live on earth. The final judgments here will be only what we know, and just rewards are provided to those who have every reason to expect them. We all are released into oneness at the very end, but the journey could seem eternal.

The question also arises, does it justify suicide if our last experience might not be so unpleasant? There are many forms of suicide which are gentle, but that is not why religions generally argue against it. The moods which lead to suicide, with the exception of the terminally ill, usually build up over a period of time: deep feelings of helplessness and pain that would not be fun to relive for eternity. This is probably why all orthodox religions forbid suicide: depression and anger may pass, but our trip to forever never ends. It could be hellish if we don't work it out before we leave. Our human lifetime is the one chance we have to make sure any future lifetimes will be the nice everlasting, even if it all takes place during death.

The Universal Journey: Light unto Light

Luckily, when we take a good look at our lives, most of us are not dissatisfied. The great majority of us, therefore, can probably look forward to experiencing the blessed miracle of losing both our perception of earthly time and the discrimination of comparative thought in one smooth curve down to our last living moment. The beauty of this elegant process, described medically as normal brain death, is that we will find ourselves lost in timelessness before we get halfway to the end, and it will take forever to get us even that far. This very personal regression to the infinite is reminiscent of equally unworldly phenomena at quite the opposite end of the size scale.

In astronomy, we know of stellar objects known as "black

holes," visible only as dots of utter darkness. A collapsed star in the center has shrunken to a point so dense that its gravity lets nothing at all, even light, escape from its surface. There is an imaginary ring in space around each black hole, referred to as the Schwarzschild radius or the "event horizon". This is the outer limit of its swirling gravitational field, the dark vortex that can seize anything at all and whirl it into that darkness forever. If we are observing an object in space and it slips over that event horizon, it will not just disappear immediately from our sight. As it approaches the rim, the vortex will start to suck in any reflected light. Massive time distortions start to occur as the object spirals to the invisible center. To any observer looking through a telescope, the object will seem to reach the event horizon and freeze in place as it slowly fades away. The moment it fell over the edge, there's nothing left but the old image; everything else has already gone beyond, where human eyes cannot see. Time has essentially stopped.

If you were passing over the event horizon, observers might see you at the very edge of darkness, rather like the Cheshire Cat with a last smile, fading away forever while you, now surrounded by your own light, in another time and space, are traveling forever into the center of brightness, the final perfect union with perfect union itself, the solid, completely compressed brilliant stuff of the primeval universe. Nothing has returned from that journey; even stars wink out when they meet the event horizon. They are going to where we cannot follow, over the edge of darkness to the city of light.

Death, as we watch it occur, seems very much the same. We see our loved ones simply posed, as they were poised a moment ago, in their last visible form in this universe - while in another, deeper, reality they have already started toward a true and timeless light. We on the outside could see eons pass before their journey will end or that light could fail. They are moving now into the eternity that we left so long ago, and the journey will take them just about forever and not a moment less. The universe finally unfolds itself for us again in a profound return, like the return of the tide that sweeps us into the endless sea. It makes no difference if it takes five days or five minutes or five seconds. Time will stop for each of us.

As memories simplify we are greeted and accepted and transported backwards into places of greater and greater love, for who did not love us as infants? Back we spiral in time with the love we found here, with days, months, centuries appearing between the moments of earthly time. We begin to circle endlessly into the center of the only universe we knew, our everlasting light, our welcome home. Suns, moons, stars; all can come and go many times, and that light will never fail. We have never been far away from eternity; we carry it with us all our days. It finally comes for us when it is time for time to transcend again in the clear light of death.

We have our time here and much to do. And then, we will return home again. We never expected most of what has come to us in this life; but this is one thing we can almost surely count on if we have any faith in logic and basic neuroscience. Otherwise, we may as well put our faith in any other religious belief since even with their cultural and historical variation, they all come to nearly identical conclusions. The great prophets of the past used legend and poetry to teach us how to live well, and how to die well. We have so much more power in this age of science that it is important to know that the best science we have still tells the same story.

We are born from eternity into the heart of love, we are each absolutely unique and ultimately universal; each of us is now and each of us is forever. Christian, Muslim, and Jew can praise God for giving us such a blessed system, as well as a Prophet and a Savior to show us how it works. Hindus, Buddhists and Taoists can accept a neuro-dharma with ease. Karma is conserved, Nirvana is nearby, and going with the flow seems to take us naturally to the stars.

We become once and we never unbecome. We all experience life until we experience death, and it is death itself that will take us on our endless journey to our expected, and appointed, meeting with eternity. We are now and we are forever; we can bet on that with very good statistical probability, from everlasting to everlasting for sure. Our story will not be repeated, but it's a happy ending. We can have faith in that, and every reason to believe.